Lecture Notes in Statistics

Edited by J. Berger, S. Fienberg, J. Gani,
K. Krickeberg, and B. Singer

49

Steffen L. Lauritzen

Extremal Families and Systems of Sufficient Statistics

Springer-Verlag

New York Berlin Heidelberg London Paris Tokyo

Author

Steffen L. Lauritzen
Department of Mathematics and Computer Science
Institute of Electronic Systems, Aalborg University
Strandvejen 19, 9000 Aalborg, Denmark

Mathematics Subject Classification: 62A99, 62B05

ISBN 0-387-96872-5 Springer-Verlag New York Berlin Heidelberg
ISBN 3-540-96872-5 Springer-Verlag Berlin Heidelberg New York

Printed in Germany

Printing and binding: Druckhaus Beltz, Hemsbach/Bergstr.
2847/3140-543210

Preface

The present book is a revised and updated version of "Statistical Models as Extremal Families" from 1982. The main difference between this and the present version is that errors have been corrected, more examples have been added and the bibliography extended and updated. Still, this is not a comprehensive treatment of the area and the reader is referred to sources such as Aldous (1985), Diaconis and Freedman (1984), Koch and Spizzichino (1982) to get a more complete picture of the situation.

The revision was finished while the author was visiting Statistical Laboratory, University of Cambridge, and he is grateful to its members for their hospitality and to the Danish Natural Science Research Council for financial support.

Cambridge, March 1988

Preface to Statistical Models as Extremal Families

The present work is an attempt to collect all previous work of the author in the area, to generalise the theory and to illustrate the theoretical concepts by working out details for specific examples.

My interest in this line of research goes back to my time as a student working with Martin-Löf's notes on statistical models from 1970, working with prediction and estimation problems in stochastic processes, where the foundations of statistical theory seemed especially unsatisfactory to me.

Over the years I have benefited much from conversations and discussions with mathematicians, statisticians and probabilists throughout the world and of course mostly to my former colleagues at the Institute of Mathematical Statistics in Copenhagen, who always have been willing to discuss these matters with me, and who have encouraged me much with their continuing interest.

Søren Johansen deserves a special thanks for reading through a version of this manuscript suggesting changes, finding errors and discussing particular details with me. So does Johan Vestergaard who followed my work with remarkable interest and has given substantial constructive criticism, especially on the topics dealt with in chapter IV.

In 1977/78 I gave a course at the University of Copenhagen on these matters and I am grateful to the students for following the course with enthusiasm and giving me many suggestions for improving the theory by their questions. Much of the material in this monograph is simply a revision of my lecture notes.

Finally I am grateful to the secretaries at the Institute in Copenhagen who all in various ways have contributed to the production of the manuscript.

Aalborg, January 1982

Steffen L. Lauritzen

Table of contents

Chapter III

Chapter IV

INTRODUCTION

The point of view behind the present work is that the connection between a statistical model and a statistical analysis is a duality (in a vague sense).

In usual textbooks on mathematical statistics it is often so that the statistical model is given in advance and then various inference principles are applied to deduce the statistical analysis to be performed.

It is however possible to reverse the above procedure: given that one wants to perform a certain statistical analysis, how can this be expressed in terms of a statistical model?

In that sense we think of the statistical analysis and the statistical model as two ways of expressing the same phenomenon, rather than thinking of the model as representing an idealisation of "truth" and the statistical analysis as a method of revealing that truth to the scientist.

It is not the aim of the present work to solve the problem of giving the correct and final mathematical description of the quite complicated relation between model and analysis. We have rather restricted ourselves to describe a particular aspect of this, formulate it in mathematical terms, and then tried to make a rigorous and consequent investigation of that mathematical structure.

The elements that we have focused upon here are the following: Firstly, any statistical model or analysis involves a notion of <u>repetitions</u> of an experiment. The simplest case of this is the vague notion of independent and identical repetitions of an experiment. This seems far too simple to describe the pattern of repetition associated with any practical statistical analysis. We give here (Chapter IV) a rigorous definition of a repetitive structure, basically introduced by Martin-Löf (1974).

Secondly, what corresponds to the statistical analysis is here idealised as a system of functions defined on the sample spaces in the repetitive structure, that are to be interpreted as sufficient statistics. These are not allowed to be arbitrary but have to satisfy certain consistency conditions ensuring that the repetitive structure is preserved in a particular sense.

The characteristic feature of a sufficient statistic is that the conditional distribution of the observation given the statistic is known, i.e. specified a priori. Thus, a system of conditional distributions given the statistic is again an important part of the model specification.

We then investigate two mathematical constructions: the construction of the maximal and the extremal family corresponding to the above described specifications.

We show that these constructions are well defined and investigate their properties in detail. We show through examples that certain well known statistical models can be constructed in this way, that others cannot and give also examples of "new" statistical models corresponding to extremal families.

The extremal families are the most interesting since the maximal families are obtained from the extremal families in a very simple way (by a mixture construction) so we concentrate on investigating the extremal families.

It should be emphasized that behind our investigations is a sort of a statistical 'principle': that _statistical models used in a statistical analysis should be chosen extremal_ in a meaningful repetitive structure.

The procedure described above gives a way of 'building' a statistical model that automatically will correspond to an extremal family. But we are also interested in _modifying a given model_ such as to make it correspond to an extremal family.

By this we mean that we can embed the experiment under consideration into a suitable repetitive structure and search for the extremal family having the _same sufficient statistics and the same conditional distributions_ given the statistic as the initial model.

This procedure will lead to modifications as _conditioning procedures_ and _extensions_ of the model. This is due to the fact that the initial model then will correspond to a subfamily of the _maximal_ family and the process of changing the original family to the extremal thus involves expressing the probability measures as _mixtures_ of those in the extremal family and, if not all extreme points are represented in those mixtures, _extending_ the model such as to include them all. This modification will always be compatible with the observations if the original model is, since it only involves a change of statements relating to the _asymptotic_ (tail) behaviour when the experiment becomes infinitely large.

Note that our considerations do not have much in common with discussions among representatives of various 'schools' of statistical inference. They do not support nor oppose a Bayesian, decision theoretic, likelihood etc. point of view, but deal with intrinsic properties of the model itself.

There are some relations between this work and discussion of 'conditionality' in statistical inference. We think of the various attempts to formulate principles of the type that under certain circumstances certain inferences should be performed in a suitable conditional distribution rather than in the distributions of the model itself.

Such conditionality arguments are of the same type as those involved here since they are precisely concerned with modifications of the model (by conditioning) rather than choosing a procedure for calculating estimates etc. etc. It should however be emphasized that 'extremality' arguments also will lead to <u>conditioning on unobservable</u> (tail) <u>random variables</u>, treating these as parameters. This is different from usual conditionality arguments.

Example II 9.7 is demonstrating the familiarity between 'extremality' and 'conditionality' very clearly. We show in this example that the <u>conditional</u> (given the values of the column sums) <u>Rasch model</u> is extremal in the repetitive structure where the number of columns tend to infinity whereas their length stays fixed. This is a typical example of a nuisance parameter problem where other conditionality arguments lead to the same modification of the model.

Extremal families are <u>maximally conditional</u> and it seems conceivable that working with extremal families to some extent could make other conditionality arguments superfluous.

The main result in the general theory is contained in Theorems II 4.4 and IV 4.1 (where the latter is a more general version of the former). It is a consequence of these theorems that the unknown probability measure in an extremal family is asymptotically identifiable when the experiment becomes infinitely large. More precisely, if A is any event in a finite experiment belonging to the repetitive structure, the conditional probability $P(A|t)$ of the event given the statistic in a large experiment will, for the large experiment becoming infinitely large, converge with probability one to the 'true' unknown probability $P(A)$ of A (and $P(A|t)$ is known since t is sufficient). (In fact $P(A|t)$ is not only strongly consistent but also an unbiased estimate of $P(A)$).

Expressing this somewhat differently, leads to noting that the 'parameter' (= the unknown probability measure) has an operational meaning in the sense that it is the limit, properly interpreted, of the sufficient statistic in the infinitely large experiment (population).

This property is of the same type as ergodicity of a stationary stochastic process $(X_t, t = 0, \pm 1, \ldots)$. The probability of any event $A \subseteq \mathcal{X}$ where $\mathcal{X}$ is the state space of the process, has a 'probabilistic' meaning as the average of the indicator of that event with respect to the probability measure governing the process. If the process is ergodic this average is (almost surely) identical to the (operational) time average

$$P(A) \overset{a.s.}{=} \lim_{T\to\infty} \frac{1}{2T+1} \sum_{s=-T}^{T} 1_A(X_s)$$

where 1_A is the indicator function of the set A. The parameters of a model for such a process are thus asymptotically identifiable if the model only contains ergodic measures. And, as for extremality, ergodicity is an assumption that cannot be verified (nor the opposite) from observing a single realisation of the process.

It seems somewhat of a drawback that in many examples, the problem of actually identifying an extremal family is hard and demands an unreasonable amount of technical mathematics and tricky arguments as the reader will observe going through our examples.

The most important class of extremal families are those considered in Chapter III where we show that the extreme point problem corresponding to statistics that combine as a commutative semigroup has a neat solution showing that generalised exponential families are extremal in the repetitive structure corresponding to 'identical independent repetitions'. The result is obtained without too much work and adds yet a fundamental property to the exponential families that also in other respects play an important role in statistical theory. We have in that chapter used some effort to show that also the estimation theory in these families is simple.

The linear normal models are dealt with in Chapter IV and can also be handled with some ease. It does not seem to be unreasonable to state that simple algebraic properties of the sufficient statistics give rise to simple solutions for the extremal family.

For many other examples the solutions are complicated and so are the methods of deriving these solutions. But maybe the complexity of the solutions simply reflect a genuine complexity in the models investigated that it is worthwile having in mind.

From the point of view of novelty and importance for statistical work the models for complex repetitive structures seem to be most interesting, the Rasch model in various versions being a main example in our treatment. It is not unexpected that the investigation of such models lead to complex problems.

The mathematical level is off and on rather advanced. We have tried to make chapters I, II and III almost self contained and quite detailed. These chapters deal with the discrete case and all the basic ideas and constructions are demonstrated here. In the concrete examples we felt it necessary to rely on quite many results from probability theory and the sometimes technical and hard proofs here could probably not be read easily by readers without a fairly good background in this area.

In chapter IV we formulate the general theory and use results from measure theory, functional analysis etc. to be able to fight successfully with the various technicalities. Familiarity with Choquet's theorem (in the metrisable version) about representation of convex simplices and its applications is not an unimportant prerequisite for the reader to grasp all parts of this work.

Chapter I

THE CASE OF A SINGLE EXPERIMENT AND FINITE SAMPLE SPACE

1. Basic facts. Maximal and extremal families

This chapter contains some very simple but basic facts concerning the relationship between a statistical model (a family of probability distributions) and a sufficient statistic.

The results are all more or less well known although seldom explicitly stated. All proofs are based on elementary methods only and are given in detail although the results could be obtained from standard theorems about sufficient statistics and subfields. We think however that it is of interest to see how easily all the results can be derived directly, especially because we later shall need fairly advanced mathematics to give analogues of these results in cases that are of more direct interest to us than the present.

In this chapter there is very little structure and thus also few results. Nevertheless these results are fundamental to the theory.

Let $\mathcal{X},\mathcal{Y}$ be finite sets. $\mathbb{P}(\mathcal{X})$ resp. $\mathbb{P}(\mathcal{Y})$ denote all probability measures on $\mathcal{X}$ resp. $\mathcal{Y}$.

Sometimes X will denote a random variable taking values in $\mathcal{X}$ with distribution $P\in\mathbb{P}(\mathcal{X})$ and we shall write for $x\in\mathcal{X}$.

$$P(\{x\})=P(x)=P\{X=x\}$$

Let $t:\mathcal{X}\to\mathcal{Y}$ be a mapping. t induces a mapping Π^t:

$$\Pi^t:\mathbb{P}(\mathcal{X})\to\mathbb{P}(\mathcal{Y})$$

given as

$$\Pi^t(P)(y)=\sum_{\{x:t(x)=y\}} P(x)=P\{t^{-1}(y)\}.$$

We shall occasionally use the notation

$$\mathcal{X}^y=\{x:t(x)=y\}.$$

For convenience we shall assume that t is <u>surjective</u>, i.e. that $t(\mathcal{X})=\mathcal{Y}$. This implies that also Π^t is surjective.

If $\mathcal{P}\subseteq\mathbb{P}(\mathcal{X})$ is a family of probability measures on $\mathcal{X}$, we say as usual that t is <u>sufficient</u> if there exists a family

$$\{Q^y, y\in\mathcal{Y}\}, \text{ with } Q^y\in\mathbb{P}(\mathcal{X})$$

and support $\text{supp}Q^y\subseteq\mathcal{X}^y$, i.e.

$$Q^y(x)=0 \text{ if } x\notin\mathcal{X}^y$$

such that for all $P\in\mathcal{P}, x\in\mathcal{X}$

$$P(x)=Q^{t(x)}(x)\Pi^t(P)(t(x)) \tag{1.1}$$

or equivalently that

$$P\{X=x|t(X)=y\}=Q^y(x).$$

Let us first note the following

<u>1.1 Proposition</u> <u>If t is sufficient for $\mathcal{P}$ then the restriction of Π^t to $\mathcal{P}$ is injective (one-to-one)</u>.

<u>Proof</u>

From (1.1) we get for $P_1,P_2\in\mathcal{P}$ with $\Pi^t(P_1)=\Pi^t(P_2)$:

$$\begin{aligned}P_1(x)&=Q^{t(x)}(x)\Pi^t(P_1)(t(x))\\&=Q^{t(x)}(x)\Pi^t(P_2)(t(x))\\&=P_2(x), \text{ which was to be proved. } \square\end{aligned}$$

<u>1.2 Proposition</u> <u>The formula</u>

$$Q_t(R)(x)=Q^{t(x)}(x)R(t(x)) \tag{1.2}$$

<u>defines an injective mapping</u> $Q_t:\mathbb{P}(\mathcal{Y})\to\mathbb{P}(\mathcal{X})$.

Proof First we have to show that

$$Q_t(\mathbb{P}(\mathcal{Y})) \subseteq \mathbb{P}(\mathcal{X}) .$$

Obviously $Q_t(R)(x) \geqq 0$ for all x. We get further

$$\sum_{x \in \mathcal{X}} Q_t(R)(x) = \sum_{y \in \mathcal{Y}} \sum_{x \in \mathcal{X}^y} Q_t(R)(x)$$

$$= \sum_{y \in \mathcal{Y}} \sum_{x \in \mathcal{X}^y} Q^y(x) R(y)$$

$$= \sum_{y \in \mathcal{Y}} R(y) \sum_{x \in \mathcal{X}^y} Q^y(x)$$

Now since $\operatorname{supp} Q^y \subseteq \mathcal{X}^y$ and $Q^y \in \mathbb{P}(\mathcal{X})$, the last quantity is equal to

$$\sum_{y \in \mathcal{Y}} R(y) \sum_{x \in \mathcal{X}} Q^y(x) = \sum_{y \in \mathcal{Y}} R(y) \cdot 1 = 1.$$

To see that Q_t is injective we choose for each $y \in \mathcal{Y}$ an $x^y \in \mathcal{X}^y$ such that $Q^y(x^y) > 0$ (such an x^y obviously exists). If then we have $R_1, R_2 \in \mathbb{P}(\mathcal{Y})$ and $Q_t(R_1) = Q_t(R_2)$ we get that for all $y \in \mathcal{Y}$

$$Q_t(R_1)(x^y) = Q^y(x^y) R_1(y)$$

$$= Q^y(x^y) R_2(y) .$$

Dividing by the positive quantity $Q^y(x^y)$, we get $R_1(y) = R_2(y)$ which was to be proved. □

1.3 Proposition The mapping Π^t is the left inverse to Q_t:

$$\Pi^t(Q_t(R)) = R, \quad \forall R \in \mathbb{P}(\mathcal{Y}) .$$

Proof Let $R \in \mathbb{P}(\mathcal{Y})$ and $P = Q_t(R)$. Then $P(x) = Q^{t(x)}(x) R(t(x))$ and

$$\Pi^t(Q_t(R))(y) = \Pi^t(P)(y) = \sum_{x \in \mathcal{X}^y} P(x)$$

$$= \sum_{x \in \mathcal{X}^y} Q^y(x) R(y) = R(y) \cdot 1$$

and thus $\Pi^t(Q_t(R))=R$ which was to be proved. □

In general Q_t is not the left inverse of Π^t since Π^t is not injective. We have however

1.4 Proposition The mapping Q_t is the left inverse of Π^t's restriction to $\mathcal{P}$:

$$Q_t(\Pi^t(P))=P, \quad \forall P\in\mathcal{P}. \tag{1.3}$$

Proof Writing (1.3) for all x

$$Q_t(\Pi^t(P))(x)=Q^{t(x)}(x)\Pi^t(P)(t(x))$$

we get the result directly from (1.1). □

In the special case when $\mathcal{P}=Q_t(\mathbb{P}(\mathcal{Y}))$ we have

1.5 Proposition If $\mathcal{P}=Q_t(P(\mathcal{Y}))$ then Π^t's restriction to $\mathcal{P}$:

$$\Pi^t|\mathcal{P}:\mathcal{P}\to\mathbb{P}(\mathcal{Y})$$

is bijective (one-to-one and onto), Q_t and $\Pi^t|\mathcal{P}$ are each others inverses and t is minimal sufficient.

Proof The first statement of the proposition is just a reformulation of propositions 1.3 and 1.4, taking into account that $\mathcal{P}=Q_t(P(\mathcal{Y}))$ and that $\Pi^t|\mathcal{P}$ therefore is surjective.

That t is minimal sufficient is seen in the following way: t is sufficient because if $P\in\mathcal{P}$, $P=Q_t(R)$ we have

$$P(x)=Q^{t(x)}(x)R(t(x))$$

$$=Q^{t(x)}(x)\Pi^t(P)(t(x))$$

since $\Pi^t(P)=\Pi^t(Q_t(R))=R$.

t is minimal sufficient because $\mathcal{P}\supseteq\{Q^y, y\in\mathcal{Y}\}$ and all the supports of Q^y are disjoint. □

Remark In fact it is clear that if $\mathcal{P}=Q_t(\mathrm{P}(\mathcal{Y}))$, then t is a complete sufficient statistic for $\mathcal{P}$. If we namely have that

$$E_P f(t(X))=0 \qquad \forall P\in\mathcal{P},$$

we get

$$\sum_{y\in\mathcal{Y}} f(y)R(y)=0 \qquad \forall R\in\mathbb{P}(\mathcal{Y}).$$

If we take

$$R(y)=\varepsilon_{y_0}(y)=\begin{cases}1 \text{ if } y=y_0\\ 0 \text{ otherwise,}\end{cases}$$

we have $\varepsilon_{y_0}\in\mathbb{P}(\mathcal{Y})$ for all $y_0\in\mathcal{Y}$ and thus

$$\sum_{y\in\mathcal{Y}} f(y)\varepsilon_{y_0}(y)=f(y_0)=0 \qquad \forall y_0\in\mathcal{Y},$$

and thus that t is complete. It then also follows that t is minimal because it is complete. □

1.6 Proposition *If $\mathcal{P}=Q_t(\mathbb{P}(\mathcal{Y}))$ then $\mathcal{P}$ is maximal with the property that t is sufficient for $\mathcal{P}$.*

Proof Let $P^*\in\mathbb{P}(\mathcal{X})$ and $P^*\notin\mathcal{P}$. Consider the family

$$\mathcal{P}^*=\mathcal{P}\cup\{P^*\}.$$

If t is sufficient for $\mathcal{P}^*$ then $\Pi^t|\mathcal{P}^*$ is injective. But this cannot be the case because $\Pi^t|\mathcal{P}$ is already a surjective mapping from $\mathcal{P}$ to $\mathbb{P}(\mathcal{Y})$ and there is thus a $P\in\mathcal{P}$ such that $\Pi^t(P)=\Pi^t(P^*)$. □

1.7 Definition The family $\mathcal{P}=Q_t(\mathrm{P}(\mathcal{Y}))$ is said to be *maximal with respect to* t.

With this definition we thus have that if $\mathcal{P}$ is maximal w.r.t. t, then t is complete and minimal sufficient for $\mathcal{P}$.

For any fixed t there are obviously different maximal families corresponding to different systems $\{Q^y, y \in \mathcal{Y}\}$. With some modification we can say that t gives a decomposition of $\mathbb{P}(\mathcal{X})$ into basically disjoint maximal subfamilies.

More precisely, let $\mathbb{P}(\mathcal{X})^+$ be the set of probabilities in $\mathbb{P}(\mathcal{X})$ with $P(x)>0$ for all $x \in \mathcal{X}$. Similarly, for a family $\mathcal{P}$ we let

$$\mathcal{P}^+ = \mathcal{P} \cap \mathbb{P}(\mathcal{X})^+.$$

We then have

<u>1.8 Proposition</u> <u>If</u> $\mathcal{P}_1$ <u>and</u> $\mathcal{P}_2$ <u>are different maximal families with respect to</u> t, <u>then</u>

$$\mathcal{P}_1^+ \cap \mathcal{P}_2^+ = \emptyset.$$

<u>Furthermore</u>

$$\mathbb{P}(\mathcal{X}) = \bigcup_{\substack{\mathcal{P} \text{ maximal} \\ \text{w.r.t. } t}} \mathcal{P} .$$

<u>In other words, each</u> $P \in \mathbb{P}(\mathcal{X})$ <u>is a member of at least one family that is maximal w.r.t.</u> t. <u>If</u> $P \in \mathbb{P}(\mathcal{X})^+$ <u>it is a member of exactly one</u>.

<u>Proof</u> Let $P \in \mathbb{P}(\mathcal{X})$. For all y that are such that $\Pi^t(P)(y)>0$ we define

$$Q^y(x) = \frac{P(x)}{\Pi^t(P)(y)} 1_{\mathcal{X}^y}(x),$$

where

$$1_A(x) = \begin{cases} 1 \text{ for } x \in A \\ 0 \text{ for } x \notin A. \end{cases}$$

For those y with $\Pi^t(P)(y)=0$ we can for example choose Q^y as

$$Q^y(x) = \frac{1_{X^y}(x)}{|X^y|},$$

where

$|A|$ = the number of elements in A.

Defining $\{Q^y, y \in \mathcal{Y}\}$ as above it is obvious that

$$P = Q_t(\Pi^t(P))$$

and thus that $P \in Q_t(\mathbb{P}(\mathcal{Y}))$, a maximal family.

If $P_1, P_2 \in \mathbb{P}(\mathcal{X})^+, P_1 \in \mathcal{P}_1, P_2 \in \mathcal{P}_2, \mathcal{P}_1 \neq \mathcal{P}_2$, we have that their conditional probabilities are uniquely defined and equal to

$$Q_1^y(x) = \frac{P_1(x)}{\Pi^t(P_1)(y)} 1_{X^y}(x), \quad Q_2^y(x) = \frac{P_2(x)}{\Pi^t(P_2)(y)} 1_{X^y}(x).$$

Thus we must have

$$\mathcal{P}^1 = Q_t^1(\mathbb{P}(\mathcal{Y})), \quad \mathcal{P}_2 = Q_t^2(\mathbb{P}(\mathcal{Y})).$$

If $Q_1^y(x) = Q_2^y(x)$ for all x,y, $\mathcal{P}_1 = \mathcal{P}_2$, but we assumed $\mathcal{P}_1 \neq \mathcal{P}_2$.

If $Q_1^y(x) \neq Q_2^y(x)$ we obviously have $P_1 \notin Q_t^2(\mathbb{P}(\mathcal{Y})) = \mathcal{P}_2$ and thus

$$\mathcal{P}_1^+ \cap \mathcal{P}_2 = \emptyset,$$

which implies the result. □

If $\mathcal{P} = Q_t(\mathbb{P}(\mathcal{Y}))$ is a maximal family $\mathcal{P}$ can be injectively parametrized by $\mathbb{P}(\mathcal{Y})$. In fact we write for $\theta \in \Theta = \mathbb{P}(\mathcal{Y})$

$$P_\theta(x) = Q_t(\theta)(x) = Q^{t(x)}(x)\theta(t(x)).$$

__1.9 Proposition__ _If_ $P=\{P_\theta,\theta\in\Theta\}=\{Q_t(\theta),\theta\in\Theta=\mathbb{P}(\mathcal{Y})\}$, _the maximum likelihood estimate_ $\hat{\theta}_x$ _of_ θ _when observing_ x _is given by_

$$\hat{\theta}_x = \varepsilon_{t(x)}. \tag{1.4}$$

__Proof__ The likelihoodfunction is equal to

$$L(\theta,x) = P_\theta(x) = Q^{t(x)}(x)\theta(t(x)).$$

which is obviously maximized for $\theta(t(x))=1$ and thus $\hat{\theta}_x=\varepsilon_{t(x)}$. □

__1.10 Corollary__ _If_ $\mathcal{P}$ _is the maximal family above, the maximum likelihood estimator_

$$\hat{\theta}:\mathcal{X}\to\Theta = \mathbb{P}(\mathcal{Y})$$

given by (1.4) _is minimal sufficient and_

$$\text{range}(\hat{\theta}) = \hat{\theta}(\mathcal{X}) = \{\varepsilon_y, y\in\mathcal{Y}\}$$

__Proof__ Proposition 1.9 shows that the maximum likelihood estimate is in a one-to-one correspondence with t by

$$\hat{\theta}_x = \varepsilon_{t(x)}\leftrightarrow t(x),$$

and the statement follows. □

We shall also consider the family

$$\mathcal{E}(\mathcal{P}) = \{Q^y, y\in\mathcal{Y}\} = \{P_\theta, \theta=\varepsilon_y, y\in\mathcal{Y}\}.$$

__1.11 Definition__ The family $\mathcal{E}(\mathcal{P})$ is said to be _extremal with respect to_ t.

If we let $\Theta^*=\{\varepsilon_y, y\in\mathcal{Y}\}$, we have that

$$\mathcal{E}(\mathcal{P}) = \{P_\theta, \theta\in\Theta^*\}$$

and

__1.12 Corollary__ _In_ $\mathcal{E}(\mathcal{P})$ _the maximum likelihood estimate is given by_

$$\hat{\theta}_x = \varepsilon_{t(x)}.$$

Proof Identical to that of prop. 1.9. □

1.13 Proposition In $\mathcal{E}(\mathcal{P})$ the maximum likelihood estimator

$$\hat{\theta}: X \to \Theta^*$$

is minimal sufficient and surjective.

1.14 Corollary t is minimal sufficient for $\mathcal{E}(\mathcal{P})$.

Proof (of prop. 1.13 and cor. 1.14): $\hat{\theta}$ is surjective because t is. t is minimal sufficient simply because all the supports of Q^y are disjoint. $\hat{\theta}$ is minimal sufficient because it is in a one-to-one correspondence with t. □

Remark Also in this case t is complete sufficient by the same argument as in the remark to prop. 1.5.

Finally we have the following result giving a special characterisation of the points in $\mathcal{E}(\mathcal{P})$ among those in $\mathcal{P}$.

1.15 Proposition Let $\mathcal{P}$ be a maximal family with respect to t. An element $P \in \mathcal{P}$ is in $\mathcal{E}(\mathcal{P})$ if and only if the distribution of $t(X)$, when X has distribution P, is degenerate.

Proof Immediate from the definition of $\mathcal{E}(\mathcal{P})$. □

Thus we see that the elements of $\mathcal{E}(\mathcal{P})$ have special properties: they maximise the likelihoodfunction, they have degenerate distribution of the statistics and they define the mapping Q_t determining $\mathcal{P}$.

The maximal and extremal families both have the property that t is complete and minimal sufficient for $\mathcal{P}$ or $\mathcal{E}(\mathcal{P})$.

The maximal families are characterised by the fact that the value of the sufficient statistic t(X) can have all possible distributions.

2. Induced maximal and extremal families

The results in the preceding section indicate that it is to some extent possible to turn around a classical problem. Instead of asking: given a family $\mathcal{P}$ of probability distributions, what is the minimal sufficient statistic, we could ask: given a statistic t, what families $\mathcal{P}$ have t as minimal sufficient statistic?

Suppose that we have finite spaces $\mathcal{X}$ and $\mathcal{Y}$ as before and a surjective mapping $t:\mathcal{X}\to\mathcal{Y}$.

Let there also be given a system $\{Q^y, y\in\mathcal{Y}\}$ of distributions on $\mathcal{X}$ with

$$\text{supp } Q^y \subseteq \mathcal{X}^y = \{x:t(x)=y\}.$$

Q^y defines as before the mapping Q_t from $\mathbb{P}(\mathcal{Y})$ to $\mathbb{P}(\mathcal{X})$. We define:

<u>2.1 Definition</u> The family

$$\mathcal{P} = Q_t(\mathrm{P}(\mathcal{Y}))$$

is the <u>maximal family</u>, and

$$\mathcal{E}(\mathcal{P}) = \{Q^y, y\in\mathcal{Y}\}$$

the <u>extremal family, induced by</u> t <u>and</u> $\{Q^y, y\in\mathcal{Y}\}$.

The most direct way of inducing a family is that corresponding to definition 2.1, but it is possible to induce families by other means.

Suppose $\mathcal{P}_0\subseteq\mathbb{P}(\mathcal{X})$ is some family of probability measures, in general not assumed maximal nor extremal.

There is then always a well defined minimal, sufficient statistic $t:\mathcal{X}\to\mathcal{Y}$ with $\Pi^t(P)(y)>0$ for some $P\in\mathcal{P}_0$. Any statistic defining the same partition of $\mathcal{X}$ as the likelihood function will do. We can then define for $y\in\mathcal{Y}$

$$Q^y(x) = \frac{P(x)}{\Pi^t(P)(y)} 1_{X^y}(x) \tag{2.1}$$

where this expression does not depend on $P \in \mathcal{P}_0$. Using the system (2.1) and the method of inducing described by definition 2.1, we get

2.2 Definition The maximal resp. extremal family induced by $\mathcal{P}_0$ is the maximal resp. extremal family induced by t and the system (2.1), where t is minimal sufficient for $\mathcal{P}_0$.

Remark Since a minimal sufficient statistic is unique up to bijective transformations ($\mathcal{P}_0$-almost surely) the definition does not depend on any particular choice of t.

The maximal family induced by $\mathcal{P}_0$ is in general an extension of $\mathcal{P}$, whereas the extremal family is not.

We shall sometimes be interested in inducing statistical models directly from the statistics alone.

As we see from proposition 1.8, there are many families that are maximal with respect to t: together they constitute the entire $\mathbb{P}(\mathcal{X})$.

What we need is a way to choose the system $\{Q^y, y \in \mathcal{Y}\}$. If we have no other structure to take into account, the most natural choice seems to be to take Q^y to be uniform on the sets $\mathcal{X}^y$:

$$Q^y(x) = \frac{1_{\mathcal{X}^y}(x)}{|\mathcal{X}^y|} . \tag{2.2}$$

Borrowing a terminology from statistical mechanics we shall define

<u>2.3 Definition</u> The system of distributions (2.2) is called the <u>microcanonical distributions associated with</u> t. The function

$$f_t(y) = |X^y|$$

is called <u>the structure function</u> of t.

Now we can induce families from a statistic t alone by using t and the microcanonical distributions in connection with definition 2.1.

<u>2.4 Definition</u> <u>The maximal family induced by</u> t is that induced by t and the microcanonical distributions. <u>The extremal family induced by</u> t is the family of microcanonical distributions.

We have thus three basically different situations in which we can induce maximal and extremal families:

1) $t:X\to Y$ and $\{Q^y, y\in Y\}\subseteq \mathbb{P}(X)$ given;

2) A family $P_0\subseteq \mathbb{P}(X)$ given;

3) A statistic $t:X\to Y$ given.

We shall later discuss the relevance of these possibilities.

3. Convexity, maximal and extremal families

We can in a natural way think of $\mathbb{P}(\mathcal{X})$ as a convex subset of $\mathbb{R}^{\mathcal{X}}$ by the embedding

$$P \leftrightarrow (P(x), x \in \mathcal{X}).$$

Doing that we shall still use the notation $\mathbb{P}(\mathcal{X})$ for that convex subset.

If $t: \mathcal{X} \to \mathcal{Y}$ is a statistic the mapping $\Pi^t: \mathbb{P}(\mathcal{X}) \to \mathbb{P}(\mathcal{Y})$ can be extended to a <u>linear</u> mapping $\tilde{\Pi}^t: \mathbb{R}^{\mathcal{X}} \to \mathbb{R}^{\mathcal{Y}}$ by defining for any vector $v = (v(x), x \in \mathcal{X}) \in \mathbb{R}^{\mathcal{X}}$

$$\tilde{\Pi}^t(v)(y) = \sum_{x \in \mathcal{X}^y} v(x).$$

Similarly the mapping $Q_t: \mathbb{P}(\mathcal{Y}) \to \mathbb{P}(\mathcal{X})$ can be extended to a linear mapping $\tilde{Q}_t: \mathbb{R}^{\mathcal{Y}} \to \mathbb{R}^{\mathcal{X}}$ by letting

$$\tilde{Q}_t(w)(x) = Q^{t(x)}(x) w(t(x)).$$

$\tilde{\Pi}^t$ is the left inverse to $\tilde{Q}_t$:

$$\tilde{\Pi}^t \circ \tilde{Q}_t = I_{\mathbb{R}^{\mathcal{X}}}.$$

$\mathbb{P}(\mathcal{X})$ is a <u>simplex</u> in $\mathbb{R}^{\mathcal{X}}$ since

$$\mathbb{P}(\mathcal{X}) = \operatorname{conv}\{\varepsilon_x, x \in \mathcal{X}\}.$$

If $\mathcal{P}$ is a maximal family, we have $\mathcal{P} = Q_t(\mathbb{P}(\mathcal{Y}))$ and we see that all points in $\mathcal{P}$ are convex combinations of the points $Q^y, y \in \mathcal{Y}$ since for $R \in \mathbb{P}(\mathcal{Y})$

$$Q_t(R) = \sum_{y \in \mathcal{Y}} R(y) Q^y.$$

That is so because $Q^y(x) = 0$ if y is not equal to $t(x)$ and thus

$$\sum_{y \in \mathcal{Y}} R(y) Q^y(x) = R(t(x)) Q^{t(x)}(x)$$

$$= Q_t(R)(x).$$

Since R is uniquely defined from $Q_t(R)$, we have

3.1 Proposition A maximal family is a simplex in $\mathbb{R}^X$.

The Proof is already given. □

And further

3.2 Proposition The extremal family $E(P)$ consists of the corners of the simplex P:

$$P = \text{conv } E(P)$$

Proof The proposition is just a rephrasing of the remarks just before proposition 3.1. □

3.3 Proposition All points in a maximal family P have a unique representation as convex combinations of elements in the corresponding extremal family.

Proof This is an immediate consequence of proposition 3.2. and the fact that the mapping Q_t is injective □

Remark Note that the unique convex combination in prop. 3.3 is the distribution of the sufficient statistic t.

Finally we notice that the mapping $Q_t:\mathbb{P}(Y)\to\mathbb{P}(X)$, if we think of these as convex subsets of $\mathbb{R}^Y$ and $\mathbb{R}^X$ respectively, is an embedding of the simplex $\mathbb{P}(Y)$ into the simplex $\mathbb{P}(X)$, since its image is a maximal family and thus a simplex and since there is, a one-to-one correspondence between the corners $\varepsilon_y, y\in Y$ of the simplex $\mathbb{P}(Y)$ and the corners $Q^y, y\in Y$ of the maximal family.

When dealing with more general situations, $\mathbb{P}(X)$, $\mathbb{P}(Y)$, P and their analogues will also be simplices but of infinite dimension. To get similar results as above, in fact even to show that the analogues of these concepts are well defined, we will need general results about representation of convex sets in locally convex vector spaces. As one might even hint from the results and proofs given here, the fact that the maximal family $P=Q_t(\mathbb{P}(Y))$ is the corner stone of all the results, and the representation of the points in P as convex combinations of points in $E(P)$ is therefore more deeply related to our theory than just as a mathematical tool.

4. Some Examples

4.1 Example

Let $\mathcal{X}=\{0,1\}^2$, $\mathcal{Y}=\{0,1,2\}$

$$t(x_1,x_2) = x_1+x_2.$$

Let

$$q^0 = \varepsilon_{(0,0)}$$
$$q^2 = \varepsilon_{(1,1)}$$

$$q^1(x_1,x_2) = \begin{cases} p & \text{for } (x_1,x_2)=(1,0) \\ 1-p & \text{for } (x_1,x_2)=(0,1) \end{cases}$$

where $0 \leq p \leq 1$.

The maximal family $\mathcal{P}$ induced by q and t is given by

$$P \in \mathcal{P} \Leftrightarrow (1-p)P((1,0)) = p\,P((0,1)) \tag{4.1}$$

because if $R \in \mathbb{P}(\mathcal{Y})$ we have

$$Q_t(R)((1,0)) = p\,R(1) = P((1,0))$$
$$Q_t(R)((0,1)) = (1-p)R(1) = P((0,1))$$

such that (4.1) has to be fulfilled. On the other hand it is clear that since R(0), R(1), R(2) can be anything if they are nonnegative and add to 1, (4.1) is the only restriction.

If we represent the probabilities in $\mathbb{P}(\mathcal{X})$ by a tetrahedron, where the points in the tetrahedron mean probability distributions with that point as gravity center, we get the following picture:

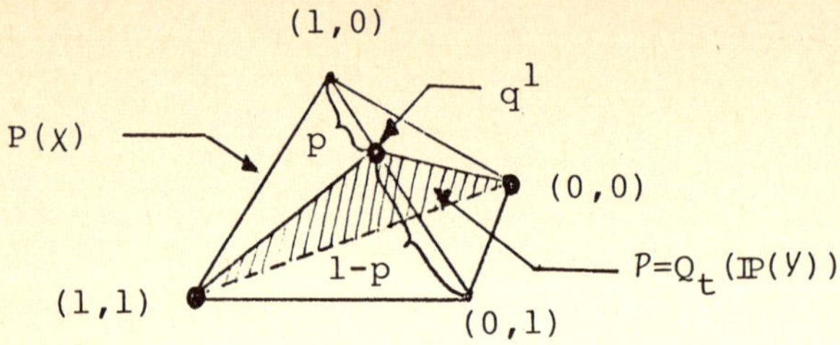

Fig. 4.1

Fig. 4.1 clearly shows how the simplex $\mathbb{P}(Y)$ is embedded into $\mathbb{P}(X)$ by Q_t.

For various values of p we get the different maximal families corresponding to t. We see as in proposition 1.8 that these together constitute a decomposition of $\mathbb{P}(X)$ and that they are not disjoint since distributions that are concentrated on $\{(0,0),(1,1)\}$ are members of all the maximal families.

The extremal family induced by t and q consists of the three points that are marked with a fat dot.

For $p=\frac{1}{2}$ we get the maximal and extremal families induced by t alone since then q^0, q^1 and q^2 are the microcanonical distributions.

4.2 Example Let X be as in example 4.1, but let $Y=\{0,1\}$ and

$$t(x_1,x_2) = x_1x_2$$

$$q^1 = \varepsilon_{(1,1)} \qquad q^0(x_1,x_2) = p_{x_1x_2},$$

where $p_{x_1x_2} \geq 0 \qquad p_{00}+p_{10}+p_{01} = 1.$

The characterisation of $\mathcal{P}$, the maximal family, is here as follows

$$P\in\mathcal{P} \Leftrightarrow P((0,0))p_{10}p_{01} = P((1,0))p_{00}p_{01}$$

$$= P((0,1))p_{00}p_{10},$$

which is seen as in the preceding example.

Using the tetrahedron again we get the following picture:

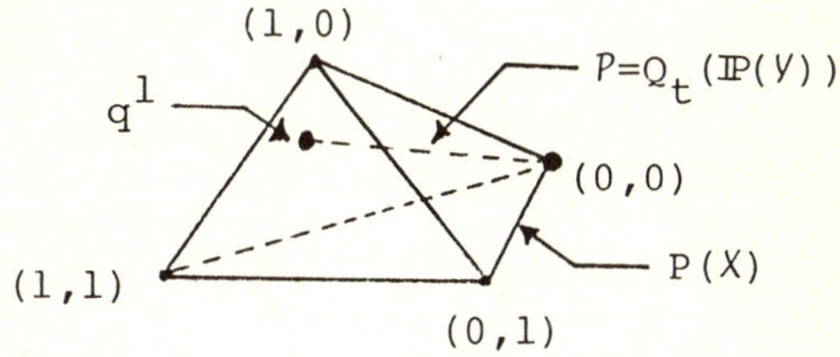

Fig. 4.2

For various values of (p_{00},p_{01},p_{10}) we get different maximal families. Their intersection is $\{\varepsilon_{(0,0)}\}$ but otherwise they are disjoint.

The extremal family induced by t and q has two points. For $p_{00}=p_{01}=p_{10}=\frac{1}{3}$ the distribution q^1 is microcanonical and we thus get the maximal resp. extremal family induced by t alone.

<u>4.3 Example</u>

Let $\mathcal{X}=\{0,1\}^n$ and consider the family of distributions P_θ, $\theta\in[0,1]$ on $\mathcal{X}$ given as

$$P_\theta(x_1,\ldots,x_n) = \theta^{x_1+\ldots+x_n}(1-\theta)^{n-(x_1+\ldots+x_n)} \tag{4.2}$$

Clearly, the mapping $t:\mathcal{X}\to\mathcal{Y}=\{0,1,\ldots,n\}$ given by

$$t(x_1,\ldots,x_n) = x_1+\ldots+x_n$$

is minimal sufficient for θ, and

$$P_\theta\{X_i=x_i,i=1,\ldots,n \mid t(X_1,\ldots,X_n)=y\}$$

$$= \frac{1}{\binom{n}{y}} 1_{\{x_1+\ldots+x_n=y\}}(x_1,\ldots,x_n), \tag{4.3}$$

i.e. the microcanonical distributions corresponding to t.

The maximal family $\mathcal{P}$ induced by $\{P_\theta,\theta\in[0,1]\}$ is thus that induced by t only and given as

$$P\in\mathcal{P} \Leftrightarrow \exists R\in\mathbb{P}(\mathcal{Y}):$$

$$P(x_1,\ldots,x_n) = \frac{R(x_1+\ldots+x_n)}{\binom{n}{x_1+\ldots+x_n}} \,. \tag{4.4}$$

This family is larger than the one we started with.

The extremal family is the family of microcanonical distributions, i.e. those with $R=\varepsilon_y$ for some $y\in\{0,1,\ldots,n\}$ in (4.4).

Thus the original family of distributions corresponding to independent identically distributed Bernoulli trials is <u>neither extremal nor maximal in this</u> (trivial) <u>repetitive structure</u>.

What we have not utilised, is the very basic fact that $x_1+\ldots+x_n$ is minimal sufficient for <u>all n</u>. It follows from the results in chapter III that the family of distributions given by (4.2) <u>is</u> extremal with the property that (4.3) holds <u>for all n</u>.

Chapter II

SIMPLE REPETITIVE STRUCTURES OF PRODUCT TYPE.DISCRETE SAMPLE SPACES

0. Conditional independence

We shall in later sections deduce a number of results that are simple consequences of certain basic properties of the notion of conditional independence.

In the present section we shall state all these results without proofs, but all proofs can readily be made with trivial manipulations.

X, Y and Z are discrete-valued random variables defined on the same probability space and we shall for the sake of brevity write $p(x|y)$ for $P\{X=x|Y=y\}$ etc. All relations between conditional probabilities shall be read with the phrase "whenever these are well-defined" following, meaning that $p(x|y)$ is well-defined iff $p(y)>0$.

<u>0.1 Definition</u> X and Z are said to be <u>conditionally independent given</u> Y, and we write

$$X \overset{Y}{\perp}_{P} Z$$

if for all y $p(x,z|y)=p(x|y)p(z|y)$ (whenever $p(y)>0$).

(0.2) $$X \overset{Y}{\perp}_{P} Z \Leftrightarrow p(x|y,z)=p(x|y)$$

(0.3) $$X \overset{Y}{\perp}_{P} Z \Leftrightarrow p(x,y|z)=p(x|y)p(y|z)$$

(0.4) $$X \overset{Y}{\perp}_{P} Z \Rightarrow X \overset{(Y,t(X))}{\perp}_{P} Z$$

(0.5) $$X \overset{Y}{\perp}_{P} Z \Rightarrow t(X,Y) \overset{Y}{\perp}_{P} Z$$

(0.6) $$X \overset{t(X)}{\perp}_{P} Y \Leftrightarrow p(y|x)=p(y|t(x))$$

$$X \perp_P^{t(X)} Y \Leftrightarrow p(x|y) = p(x|t(x))p(t(x)|y) \tag{0.7}$$

$$X \perp_P^{t(X)} Y \Leftrightarrow p(x,y) = p(x|t(x))p(t(x),y) \tag{0.8}$$

$$X \perp_P^{Y} Z \wedge (X,Y) \perp_P^{Z} W \Rightarrow X \perp_P^{Y} (Z,W) \tag{0.9}$$

1. Preliminaries. Notation

In this chapter we shall study the connection between a statistical model and corresponding sufficient statistics in situations where there are repetitions of the experiments (or rather extensions) and where these have an especially simple structure.

More precisely we are concerned with a sequence of discrete, at most denumerable (sample) spaces $X_1, X_2, \ldots$. Points in X_n are denoted by x_n.

We shall furthermore be interested in the products

$$\underset{\sim}{X}_n = X_1 \times \ldots \times X_n$$

with typical points

$$\underset{\sim}{x}_n = (x_1, \ldots, x_n).$$

We think of the family of sample spaces $(\underset{\sim}{X}_n)_{n \in \mathbb{N}}$ as the family of possible "experiments" and we shall investigate these simultaneously and take advantage of the fact that these are mutually related in that

$$\underset{\sim}{X}_{n+1} = \underset{\sim}{X}_n \times X_{n+1}.$$

It is convenient for us also to use a special notation for the products

$$_m\underset{\sim}{X}_n = X_{m+1} \times \ldots \times X_n \quad \text{for } m < n$$

$$\underset{\sim}{X} = X_1 \times \ldots \times X_n \times \ldots$$

such that

$$\underset{\sim}{X}_n = {}_0\underset{\sim}{X}_n \quad \text{and} \quad \underset{\sim}{X} = \underset{\sim}{X}_\infty = {}_0\underset{\sim}{X}_\infty .$$

Typical points in these spaces are denoted by ${}_m\underset{\sim}{x}_n$, $\underset{\sim}{x}$ etc.

The spaces are equipped with the product topology and $\mathbb{P}(\underset{\sim}{X})$, $\mathbb{P}(\underset{\sim}{X}_n)$, $\mathbb{P}({}_m\underset{\sim}{X}_n)$, $\mathbb{P}(X_n)$ etc. each denote the set of all Radon probability measures on the respective space.

The spaces are connected by <u>projections</u> and here we use the following notation for $m \leqq n$

$$\underset{\sim}{X}_m \overset{\underset{\sim}{p}_{mn}}{\longleftarrow} \underset{\sim}{X}_n$$

$$\underset{\sim}{x}_m = \underset{\sim}{p}_{mn}(\underset{\sim}{x}_n) \overset{\underset{\sim}{p}_{mn}}{\longleftarrow} \underset{\sim}{x}_n = (x_1,\ldots,x_m,x_{m+1},\ldots,x_n) = (\underset{\sim}{x}_m,{}_m\underset{\sim}{x}_n)$$

where these clearly satisfy

$$\underset{\sim}{p}_{mn}\,\underset{\sim}{p}_{nr} = \underset{\sim}{p}_{mr} \quad \text{for } m \leqq n \leqq r.$$

We let $\underset{\sim}{p}_{n\infty} = \underset{\sim}{p}_n$ and note that the above relation also holds for $r = \infty$.

We also define ${}_m\underset{\sim}{p}_n$, p_n by

$${}_m\underset{\sim}{X}_n \overset{{}_m\underset{\sim}{p}_n}{\longleftarrow} \underset{\sim}{X} \; ; \; X_n \overset{p_n}{\longleftarrow} \underset{\sim}{X}$$

$${}_m\underset{\sim}{x}_n \overset{{}_m\underset{\sim}{p}_n}{\longleftarrow} \underset{\sim}{x} \; ; \; x_n \overset{p_n}{\longleftarrow} \underset{\sim}{x} .$$

Any member of $\mathbb{P}(\underset{\sim}{X})$ is uniquely determined by its "consistent family" i.e. by all the marginal distributions

$$\Pi^{\underset{\sim}{p}_n}(P) \in \mathbb{P}(\underset{\sim}{X}_n)$$

$$\Pi^{\underset{\sim}{p}_n}(P)(A) = P(A \times {}_n\underset{\sim}{X}_\infty) .$$

We shall be interested in certain random variables on $\underset{\sim}{X}$ defined below by these projections

$$X_n(\underset{\sim}{x}) = p_n(\underset{\sim}{x}) = x_n$$

$$\underset{\sim}{X}_n(\underset{\sim}{x}) = \underset{\sim}{p}_n(\underset{\sim}{x}) = \underset{\sim}{x}_n$$

$${}_m\underset{\sim}{X}_n(\underset{\sim}{x}) = {}_m\underset{\sim}{p}_n(\underset{\sim}{x}) = {}_m\underset{\sim}{x}_n ,$$

and for a given $P \in \mathbb{P}(\underset{\sim}{X})$ their distribution is given by $\Pi^{p_n}(P)$, $\Pi^{\underset{\sim}{p}_n}(P)$ and $\Pi^{{}_m\underset{\sim}{p}_n}(P)$ resp.

Statistical models for such systems are given by a specification of a subfamily $\mathcal{P} \subseteq \mathbb{P}(\underset{\sim}{X})$ and the corresponding families of marginal distributions are denoted by $\mathcal{P}_n \subseteq \mathbb{P}(X_n)$, $\underset{\sim}{\mathcal{P}}_n \subseteq \mathbb{P}(\underset{\sim}{X}_n)$, ${}_m\underset{\sim}{\mathcal{P}}_n \subseteq \mathbb{P}({}_m\underset{\sim}{X}_n)$.

In terms of statistical functions we think of $\underset{\sim}{X}_n$ as our "observable" random variables and we are therefore interested in statistics of the form

$$t_n ; \underset{\sim}{X}_n \to Y_n ,$$

where Y_n also is discrete and at most denumerable. The statistics $\underset{\sim}{t}_n$ and $\underset{\sim}{t}$ are then defined pointwise as

$$\underset{\sim}{t}_n(\underset{\sim}{x}_n) = (t_1(x_1), t_2(\underset{\sim}{x}_2), \ldots, t_n(\underset{\sim}{x}_n)) \in \underset{\sim}{Y}_n$$

$$\underset{\sim}{t}(\underset{\sim}{x}) = (t_1(x_1), t_2(\underset{\sim}{x}_2), \ldots) \in \underset{\sim}{Y},$$

and we define corresponding random variables as

$$Y_n = t_n(\underset{\sim}{X}_n), \quad \underset{\sim}{Y}_n = \underset{\sim}{t}_n(\underset{\sim}{X}_n), \quad n \in \mathbb{N} .$$

We shall for convenience assume that the maps t_n all are <u>surjective</u>, i.e.

$$t_n(\underset{\sim}{X}_n) = \mathcal{Y}_n$$

whereas this will usually not be the case for $\underset{\sim}{t}_n$ or $\underset{\sim}{t}$.

Further we shall assume that the families we consider satisfy the property that

$$\forall n \in \mathbb{N}\ \forall y_n \in \mathcal{Y}_n\ \exists P \in \mathcal{P} : P\{Y_n = y_n\} > 0.$$

This assumption is not a severe restriction.

We shall think of the spaces $(\underset{\sim}{X}_n)_{n \in \mathbb{N}}$ and the way they are related to each other (the projections $\underset{\sim}{p}_{mn}$) as a "repetitive structure" and give a precise definition later.

2. Notions of sufficiency

The usual notion of sufficiency is defined for one single sample space X as in chapter I and does not take into account the presence of a repetitive structure. It is therefore as such not adequate (detailed enough) to make our further investigations fruitful.

We shall here discuss three notions that are relevant for us as strengthened forms of sufficiency. They are all related but different in general. In later sections we shall see how they all become relevant.

We use the notation introduced in the previous section and focus on a family of probability measures $P \subseteq \mathbb{P}(\underset{\sim}{X})$ and a statistic $\underset{\sim}{t}$ (i.e. a sequence of statistics t_n).

2.1 Definition The sequence of statistics $\underset{\sim}{t}$ is said to be transitive in P if for all $n \in \mathbb{N}$, $P \in P$, $\underset{\sim}{X}_n$ and $Y_{n+1} = t_{n+1}(\underset{\sim}{X}_{n+1})$ are conditionally independent given $Y_n = t_n(\underset{\sim}{X}_n)$, i.e.

$$\forall n \in \mathbb{N} \;\; \forall P \in P : \underset{\sim}{X}_n \overset{t_n(\underset{\sim}{X}_n)}{\perp_P} t_{n+1}(\underset{\sim}{X}_{n+1}).$$

If also t_n is sufficient for $\underset{\sim}{P}_n$ for all n, we say that $\underset{\sim}{t}$ is sufficient and transitive.

By the property (0.6) of conditional independence we see that this "means" that $t_n(\underset{\sim}{X}_n)$ carries all information in $\underset{\sim}{X}_n$ needed to predict the value of the next statistic $t_{n+1}(\underset{\sim}{X}_{n+1})$.

The transitivity is closely connected to the Markov property by the following:

<u>2.2 Proposition</u> If $\underset{\sim}{t}$ <u>is transitive, then the stochastic process</u> $(Y_1, Y_2, \ldots)$ <u>is a Markov process for all</u> $P \in \mathcal{P}$.

<u>Proof</u> That the Y-process is Markov means that for all $n \in \mathbb{N}$

$$\underset{\sim}{Y}_{n-1} \perp_P^{Y_n} Y_{n+1}.$$

Now this follows from the definition of a transitive statistic and from property (0.5) of conditional independence since

$$\underset{\sim}{Y}_{n-1} = \underset{\sim}{t}_{n-1}(p_{n-1,n}(\underset{\sim}{X}_n)). \quad \square$$

<u>Remark</u> The converse is false since the converse in (0.5) is false. Proposition 2.2 says that $\underset{\sim}{t}$ takes into account the repetetive structure in the sense that it takes any complex structure of stochastic dependence between the random variables $(X_n, n \in \mathbb{N})$ and transforms this into Markov dependence, which can be said to be the simplest thinkable structure of dependence (apart from independence). Another way of stating the same result in words is to say that the process $(\underset{\sim}{X}_n, n \in \mathbb{N})$ is always Markov and consists of observations of successively extended experiments. Demanding a statistic to be transitive implies that this structure is preserved when transforming with $\underset{\sim}{t}$.

When t_n is sufficient for $\underset{\sim}{\mathcal{P}}_n$ we have a factorisation of the probabilities in $\underset{\sim}{\mathcal{P}}$ in the sense that there is a family $Q_n^{y_n} \in P(\underset{\sim}{X}_n)$ with support in $t_n^{-1}(y_n)$ such that

$$P\{\underset{\sim}{X}_n = \underset{\sim}{x}_n\} = Q_n^{t_n(\underset{\sim}{x}_n)}(\underset{\sim}{x}_n)\ \Pi^{t_n \circ \underset{\sim}{p}_n}(P)(\underset{\sim}{x}_n). \tag{2.1}$$

The extra condition of transitivity can be formulated in terms of relations between these $Q_n^{y_n}, n \in \mathbb{N}, y_n \in \mathcal{Y}_n$:

2.3 Proposition A sufficient sequence $\underset{\sim}{t}$ is transitive iff there are conditional distributions $Q_n^{y_n}$ of $\underset{\sim}{X}_n$ given Y_n satisfying

$$Q_{n+1}^{y_{n+1}}\{\underset{\sim}{X}_n = \underset{\sim}{x}_n\} = Q_n^{t_n(\underset{\sim}{x}_n)}(\underset{\sim}{x}_n)Q_{n+1}^{y_{n+1}}\{Y_n = t_n(\underset{\sim}{x}_n)\} \tag{2.2}$$

Proof (2.2) is equivalent to the statement

$$p(\underset{\sim}{x}_n|y_{n+1}) = p(\underset{\sim}{x}_n|t_n(\underset{\sim}{x}_n))p(t_n(\underset{\sim}{x}_n)|y_{n+1})$$

in the notation of section 0. But this is just property (0.7) of conditional independence. □

Remark (2.2) simply expresses the fact that t_n is sufficient in the family of conditional distributions of X_n given Y_{n+1}.

In chapter I we saw that (2.1) could be formulated in terms of the mappings

$$\Pi^{t_n} : \underset{\sim}{P}_n \to \mathbb{P}(\mathcal{Y}_n),\ Q_{t_n} : \mathbb{P}(\mathcal{Y}_n) \to \mathbb{P}(\underset{\sim}{\mathcal{X}}_n).$$

Here we similarly get that transitivity of a sufficient sequence can be formulated in terms of the properties of these mappings

2.4 Proposition A sufficient sequence $\underset{\sim}{t}$ is transitive if and only if there are linear mappings for $m \leqq n$

$${}_mQ_n : \mathbb{P}(\mathcal{Y}_n) \to \mathbb{P}(\mathcal{Y}_m)$$

satisfying

$${}_nQ_n = I_{\mathbb{P}(\mathcal{Y}_n)},\quad {}_mQ_{n\,n}Q_r = {}_mQ_r \text{ for } m \leqq n \leqq r$$

such that the diagrams for $m \leqq n$

$$\begin{array}{ccc} \mathbb{P}(\underset{\sim}{X}_m) & \overset{\Pi^{P_{mn}}}{\longleftarrow} & \mathbb{P}(\underset{\sim}{X}_n) \\ \big\uparrow {\scriptstyle Q_{t_m}} & & \big\uparrow {\scriptstyle Q_{t_n}} \\ \mathbb{P}(\mathcal{Y}_m) & \overset{{}_m Q_n}{\longleftarrow} & \mathbb{P}(\mathcal{Y}_n) \end{array} \tag{2.3}$$

<u>all are commutative</u>.

<u>The maps</u> ${}_m Q_n$ <u>are given as</u>

$${}_m Q_n = \Pi^{t_m \circ P_{mn}} \, Q_{t_n} \, . \tag{2.4}$$

<u>Proof</u> We only have to show the commutativity for $n = m+1$ since the statement for general m follows by letting

$${}_m Q_n = {}_m Q_{m+1} {}_{m+1} Q_{m+2} \cdots {}_{n-1} Q_n$$

and using commutativity of all diagrams with $n-m = 1$.

If we rewrite (2.2) in terms of these mappings, we get

$$\Pi^{\underset{\sim}{P}_{m,m+1}} \, Q_{t_{m+1}}(\varepsilon_{Y_{m+1}}) = Q_{t_m} \, \Pi^{t_m \circ P_{m,m+1}} Q_{t_{m+1}}(\varepsilon_{Y_{m+1}})$$

If we now define

$${}_m Q_{m+1} = \Pi^{t_m \circ \underset{\sim}{P}_{m,m+1}} Q_{t_{m+1}} = \Pi^{t_m} {}_{\Pi} \underset{\sim}{P}_{m,m+1} Q_{t_{m+1}}$$

we thus have from (2.2)

$$\Pi^{\underset{\sim}{P}_{m,m+1}} Q_{t_{m+1}}(\varepsilon_{Y_{m+1}}) = Q_{t_m} {}_m Q_n(\varepsilon_{Y_{m+1}}) \, .$$

Since all involved mappings are linear they are determined from their values on $\varepsilon_{y_{m+1}}$, $y_{m+1} \in \mathcal{Y}_{m+1}$ and we have proved "only if".

If we now assume the diagrams to be commutative, we have

$$\Pi^{\rho_{mn}} Q_{t_n} = Q_{t_m} {}_mQ_n \tag{2.5}$$

and if we use the sufficiency of t_m, we get $\Pi^{t_m} Q_{t_m} = I_{\mathbb{P}(\mathcal{Y}_m)}$ and can thus apply Π^{t_m} to both sides of (2.5) which gives

$$\Pi^{t_m} \Pi^{\rho_{mn}} Q_{t_n} = {}_mQ_n$$

Which proves (2.4). If we insert that into (2.5) and use (2.5) for $m+1 = n$, we get (2.2) as in the first part of the proof. □

<u>Remark</u> Note that the condition of transitivity given here <u>does not directly involve the family</u> $\mathcal{P}$. The only way in which $\mathcal{P}$ comes into the picture is via the maps Q_{t_n}, and they do not determine the family $\mathcal{P}$ in general (they do if we demand $\mathcal{P}$ to be maximal or extremal, see next section). The maps ${}_mQ_n$ determine the conditional distribution of Y_n given Y_m, see section 3.

A similar result (weaker) can be derived for the more "direct" diagrams involving Π^{t_n} instead of Q_{t_n}.

<u>2.5 Proposition</u> <u>If $\underset{\sim}{t}$ is sufficient and transitive, then for all</u> $m \leq n$, <u>the diagrams below are commutative, where</u> ${}_mQ_n$ <u>are the mappings defined in proposition 2.4</u>.

$$
\begin{array}{ccc}
\underset{\sim}{P}_m & \overset{\Pi^{\underset{\sim}{P}_{mn}}}{\longleftarrow} & \underset{\sim}{P}_n \\
\Big\downarrow \Pi^{t_m} & & \Big\downarrow \Pi^{t_n} \\
\mathbb{P}(\mathcal{Y}_m) & \overset{{}_mQ_n}{\longleftarrow} & \mathbb{P}(\mathcal{Y}_n)
\end{array} \quad . \tag{2.6}
$$

Proof Since t_n is sufficient, we have for $\underset{\sim}{P}_n \in \underset{\sim}{P}_n$

$$\underset{\sim}{P}_n = Q_{t_n} \Pi^{t_n}(\underset{\sim}{P}_n) \ .$$

But then we get, using (2.4) that for $\underset{\sim}{P}_n \in P_n$,

$${}_mQ_n{}^{t_n}(\underset{\sim}{P}_n) = \Pi^{t_m} \Pi^{\underset{\sim}{P}_{mn}} Q_{t_n} \Pi^{t_n}(\underset{\sim}{P}_n) = \Pi^{t_m} \Pi^{\underset{\sim}{P}_{mn}}(\underset{\sim}{P}_n) \ ,$$

which was to be proved. □

Remark The converse is false. We cannot deduce commutativity of (2.3) from commutativity of (2.6) since, in general $\Pi^{t_n}(\underset{\sim}{P}_n)$ might be considerably smaller (of less dimension) than $\mathbb{P}(\mathcal{Y}_n)$.

It will be of interest to consider two other notions related to sufficiency.

2.6 Definition A sequence of statistics $\underset{\sim}{t}$ is said to be algebraically transitive if for all $n \in \mathbb{N}$

$$t_n(\underset{\sim}{x}_n) = t_n(\underset{\sim}{x}'_n) \Rightarrow t_{n+1}(\underset{\sim}{x}_n, x_{n+1}) = t_{n+1}(x'_n, x_{n+1}) \ . \tag{2.7}$$

Remark (2.7) is saying that $t_{n+1}(\underset{\sim}{x}_{n+1})$ is a function of $t_n(\underset{\sim}{x}_n)$ and x_{n+1} and is easily extended to the property

$$t_m(\underset{\sim}{x}_m) = t_m(\underset{\sim}{x}'_m) \Rightarrow t_n(\underset{\sim}{x}_m, {}_m\underset{\sim}{x}_n) = t_n(\underset{\sim}{x}'_m, {}_m\underset{\sim}{x}_n)$$

by successive use of (2.7).

Algebraic transitivity is obviously related to transitivity, but in general they do not imply each other. In words we can say that transititivity means that <u>the conditional distribution</u> of Y_{n+1} depends on $\underset{\sim}{X}_n$ only through $t_n(\underset{\sim}{X}_n)$, whereas for algebraic transitivity the <u>actual value</u> of y_{n+1} depends on $\underset{\sim}{x}_n$ only through $t_n(\underset{\sim}{x}_n)$. If $\underset{\sim}{X}_n$ and X_{n+1} were stochastically independent we would expect the latter to imply the former, which turns out to be the case.

<u>2.7 Proposition</u> <u>If for all</u> $P \in \mathcal{P}$ <u>the random variables</u> $X_1, X_2, \ldots$ <u>are mutually independent and</u> $\underset{\sim}{t}$ <u>is algebraically transitive, then</u> $\underset{\sim}{t}$ <u>is transitive</u>.

<u>Proof</u> Since $Y_{n+1} = \phi_{n+1}(Y_n, X_{n+1})$ and Y_n is a function of $\underset{\sim}{X}_n$ conditional independence to get

$$\underset{\sim}{X}_n \perp_P X_{n+1} \Rightarrow \underset{\sim}{X}_n \overset{Y_n}{\perp_P} X_{n+1}$$

and then property (0.5) implies

$$\underset{\sim}{X}_n \overset{Y_n}{\perp_P} \phi_{n+1}(Y_n, X_{n+1}) = Y_{n+1},$$

which was to be proved. □

From proposition 2.7 we can show the existence of a minimal sufficient sequence of statistics which is both transitive and algebraically transitive, (in the case where $\mathcal{P}$ consists of product measures).

If we define the normalised likelihood function as usual as

$$\bar{L}_n(x_n, P) = \frac{P(x_n)}{\bar{P}(x_n)} ,$$

if $\bar{P}(\underset{\sim}{x}_n) = \sup_{P \in \mathcal{P}} P(\underset{\sim}{x}_n) > 0$, and undefined otherwise, we get

<u>2.8 Proposition</u> <u>If for all</u> $P \in \mathcal{P}$, $X_1, \ldots, X_n, \ldots$ <u>are independent, then the normalised likelihood function is minimal sufficient, algebraically transitive and transitive</u>.

<u>Proof</u> We know that the normalised likelihood function is minimal sufficient, so all that remains to be shown is that it is algebraically transitive since the conclusion then follows from proposition 2.7.

Let $\underset{\sim}{x}_n'$ and $\underset{\sim}{x}_n''$ be such that $\bar{L}_n(\underset{\sim}{x}_n'; \cdot) = \bar{L}_n(\underset{\sim}{x}_n''; \cdot)$. We then have

$$\begin{aligned}
\bar{P}(\underset{\sim}{x}_n', x_{n+1}) &= \sup_{P \in \mathcal{P}} [P(\underset{\sim}{x}_n') P(x_{n+1})] \\
&= \sup_{P \in \mathcal{P}} [\bar{P}(\underset{\sim}{x}_n') \bar{L}_n(\underset{\sim}{x}_n', P) P(x_{n+1})] \\
&= \bar{P}(\underset{\sim}{x}_n') \sup_{P \in \mathcal{P}} [\bar{L}_n(\underset{\sim}{x}_n'', P) P(x_{n+1})] \\
&= \frac{\bar{P}(\underset{\sim}{x}_n')}{\bar{P}(\underset{\sim}{x}_n'')} \sup_{P \in \mathcal{P}} [P(\underset{\sim}{x}_n'') P(x_{n+1})] \\
&= \frac{\bar{P}(\underset{\sim}{x}_n')}{\bar{P}(\underset{\sim}{x}_n'')} \bar{P}(\underset{\sim}{x}_n'', x_{n+1})
\end{aligned}$$

from which we get

$$\frac{\bar{P}(x_n')}{\bar{P}(x_n', x_{n+1})} = \frac{\bar{P}(x_n'')}{\bar{P}(x_n'', \underset{\sim}{x}_{n+1})} \quad . \tag{2.8}$$

But then we get

$$\bar{L}_{n+1}(\underset{\sim}{x}_n', x_{n+1}; P) = \frac{P(\underset{\sim}{x}_n')P(x_{n+1})}{\bar{P}(\underset{\sim}{x}_n', x_{n+1})}$$

$$= \bar{L}_n(\underset{\sim}{x}_n'; P)P(x_{n+1})\frac{\bar{P}(\underset{\sim}{x}_n')}{\bar{P}(\underset{\sim}{x}_n', x_{n+1})}$$

$$= \bar{L}_n(\underset{\sim}{x}_n''; P)P(x_{n+1})\frac{\bar{P}(\underset{\sim}{x}_n'')}{\bar{P}(\underset{\sim}{x}_n'', x_{n+1})}$$

$$= \bar{L}_{n+1}(\underset{\sim}{x}_n'', x_{n+1}; P),$$

thus that the statistic

$$t_n(\underset{\sim}{x}_n) = \bar{L}_n(\underset{\sim}{x}_n; \cdot)$$

is algebraically transitive, which was to be proved. □

If the measures in $\mathcal{P}$ are not product measures there is no direct connection between algebraic transitivity and transitivity and $\bar{L}_n$ will neither be algebraically transitive nor transitive. In that case it seems as if we need a stronger notion of sufficiency.

The notion that seems adequate for us is that of total sufficiency which is defined as follows:

2.9 Definition The statistic $t_n : \underset{\sim}{X}_n \to \mathcal{Y}_n$ is said to be totally sufficient for $\mathcal{P}$ if

i) t_n is sufficient for $\underset{\sim}{\mathcal{P}}_n$

ii) $$\forall k \geq 1 \quad \underset{\sim}{X}_n \underset{\mathcal{P}}{\perp}^{t_n(\underset{\sim}{X}_n)} {}_n\underset{\sim}{X}_{n+k} \, .$$

Note that this definition does not involve the entire sequence $t_n, n \in \mathbb{N}$, which makes it easier to work with. This notion is also a direct strengthening of sufficiency since we have:

2.10 Proposition If for all $P \in \mathcal{P}$ $X_1, \ldots, X_n, \ldots$ are independent then t_n is totally sufficient iff it is sufficient for $\underset{\sim}{\mathcal{P}}_n$.

Proof When $\underset{\sim}{X}_n$ and ${}_n\underset{\sim}{X}_{n+k}$ are independent, condition ii) is always fulfilled. □

Remark Condition ii) ensures that $t_n(\underset{\sim}{X}_n)$ carries all information in $\underset{\sim}{X}_n$ that can be used to predicting the "future" of the stochastic process $X_1, X_2, \ldots$. It could therefore also be termed "prediction sufficient".

For totally sufficient statistics we have a simple factorisation theorem.

2.11 Proposition A statistic t_n is totally sufficient for $\mathcal{P}$ if and only if there is a factorisation of the probabilities for $k \geq 0$ as

$$P(\underset{\sim}{x}_{n+k}) = f_n(\underset{\sim}{x}_n) g_{n,k}(P, t_n(\underset{\sim}{x}_n), {}_n\underset{\sim}{x}_{n+k}) . \tag{2.9}$$

Proof Suppose t_n is totally sufficient. Then, by the conditional independence property (0.8) we have

$$P(\underset{\sim}{x}_{n+k}) = P(\underset{\sim}{x}_n | t_n(\underset{\sim}{x}_n)) P(t_n(\underset{\sim}{x}_n), {}_n\underset{\sim}{x}_{n+k}).$$

But since t_n is sufficient for $\underset{\sim}{P}_n$, we can let

$$f_n(\underset{\sim}{x}_n) = P(\underset{\sim}{x}_n | t_n(\underset{\sim}{x}_n))$$

since this does not depend on P, and

$$g_{n,k}(P, t_n(\underset{\sim}{x}_n), {}_n\underset{\sim}{x}_{n+k}) = P(t_n(\underset{\sim}{x}_n), {}_n\underset{\sim}{x}_{n+k}),$$

which gives us (2.9).

If we have the factorisation (2.9), we get

$$P(\underset{\sim}{x}_{n+k} | t_n(\underset{\sim}{x}_n), {}_n\underset{\sim}{x}_{n+k})$$

$$= \frac{f_n(\underset{\sim}{x}_n) g_{n,k}(P, t_n(\underset{\sim}{x}_n), {}_n\underset{\sim}{x}_{n+k})}{\sum\limits_{t_n(\underset{\sim}{x}'_n) = t_n(\underset{\sim}{x}_n)} f_n(\underset{\sim}{x}'_n) g_{n,k}(P, t_n(\underset{\sim}{x}_n), {}_n\underset{\sim}{x}_{n+k})}$$

$$= \frac{f_n(\underset{\sim}{x}_n)}{\Sigma f_n(\underset{\sim}{x}'_n)} = \frac{f_n(\underset{\sim}{x}_n)}{h_n(t_n(\underset{\sim}{x}_n))} = P(\underset{\sim}{x}_n | t_n(\underset{\sim}{x}_n))$$

and property (0.6) of conditional independence together with the last equality gives the result. □

The factorisation theorem has an analogous formulation

<u>2.12 Proposition</u> <u>A statistic t_n is totally sufficient if and only if there is a factorisation for $k \geq 0$ as</u>

$$P(\underset{\sim}{x}_n | {}_n\underset{\sim}{x}_{n+k}) = f_n(\underset{\sim}{x}_n) g_{n,k}(P, t_n(\underset{\sim}{x}_n), {}_n\underset{\sim}{x}_{n+k}). \qquad (2.10)$$

Proof The proof is completely analogous to that of the previous, just that we instead of property (0.8) use (0.7) of conditional independence. □

From this factorisation result we immediately deduce the following basic fact:

2.13 Proposition A statistic t_n is totally sufficient if and only if it is sufficient in the family of conditional distributions given ${}_n\underset{\sim}{X}_{n+k}$, $k \geq 1$.

Proof Since the "parameter" in that family of distributions is $(P, {}_n\underset{\sim}{x}_{n+k})$, the factorisation (2.10) is exactly the Neyman factorisation for $P(\underset{\sim}{x}_n \mid {}_n\underset{\sim}{x}_{n+k})$. The statement thus follows directly from the Neyman factorisation theorem. □

Proposition 2.13 tells us how to express total sufficiency in terms of "usual" sufficiency, namely by considering the family of all conditional distributions instead of just the marginals.

Correspondingly we should also look at another likelihood function, the normalised conditional likelihood function $\tilde{L}_n$ which we define as follows: let

$$\tilde{P}(\underset{\sim}{x}_n) = \sup_{\substack{P \in \mathcal{P} \\ k \geq 1 \\ {}_n\underset{\sim}{x}_{n+k} \in {}_n\underset{\sim}{\mathcal{X}}_{n+k}}} P(\underset{\sim}{x}_n \mid {}_n\underset{\sim}{x}_{n+k})$$

and if $\tilde{P}(\underset{\sim}{x}_n) > 0$ then

$$\tilde{L}_n(\underset{\sim}{x}_n; P, {}_n\underset{\sim}{x}_{n+k}) = \frac{P(\underset{\sim}{x}_n \mid {}_n\underset{\sim}{x}_{n+k})}{\tilde{P}(\underset{\sim}{x}_n)} .$$

We then deduce from the usual result about minimal sufficiency of the likelihood function that if we define a minimal totally sufficient statistic to be one that is a function of any other totally sufficient statistic, we have

2.14 Corollary *The normalised conditional likelihood function is minimal totally sufficient.*

Proof The result follows directly from proposition 2.13 and the usual result about minimal sufficiency of the likelihood function. □

If a sequence $\underset{\sim}{t}$ of statistics is algebraically transitive and t_n is totally sufficient for all n, we can refine the factorisation (2.9):

2.15 Proposition *If $\underset{\sim}{t}$ is algebraically transitive and t_n is totally sufficient for all n, we have the factorisation*

$$P(\underset{\sim}{x}_{n+k}) = f_n(\underset{\sim}{x}_n)g_{nk}(t_n(\underset{\sim}{x}_n),{}_n\underset{\sim}{x}_{n+k})h_{n+k}(P,t_{n+m}(\underset{\sim}{x}_{n+k})) \qquad (2.11)$$

Proof Since t_n is totally sufficient, we get from (2.9) that

$$P(\underset{\sim}{x}_{n+k}) = f_n(\underset{\sim}{x}_n)g^*_{nk}(P,t_n(\underset{\sim}{x}_n),{}_n\underset{\sim}{x}_{n+k}) \ . \qquad (2.12)$$

Because t_{n+k} is sufficient we also have

$$P(\underset{\sim}{x}_{n+k}) = g^{**}_{n+k}(\underset{\sim}{x}_{n+k})h_{n+k}(P,t_{n+k}(\underset{\sim}{x}_{n+k})) \ .$$

If $f_n(\underset{\sim}{x}_n) > 0$ this gives us an equation:

$$g^*_{nk}(P,t_n(\underset{\sim}{x}_n),{}_n\underset{\sim}{x}_{n+k}) = \frac{g^{**}_{n+k}(\underset{\sim}{x}_{n+k})}{f_n(\underset{\sim}{x}_n)}\ h_{n+k}(P,t_{n+k}(\underset{\sim}{x}_{n+k})).$$

Now since $\underset{\sim}{t}$ is algebraically transitive, t_{n+k} is a function of $t_n(\underset{\sim}{x}_n)$ and ${}_n\underset{\sim}{x}_{n+k}$ and this must thus also be the case for the ratio $g^{**}_{n+k}(\underset{\sim}{x}_{n+k})/f_n(\underset{\sim}{x}_n)$,

$$g_{nk}(t_n(\underset{\sim}{x}_n), {}_n\underset{\sim}{x}_{n+k}) = \frac{g^{**}_{n+k}(\underset{\sim}{x}_{n+k})}{f_n(\underset{\sim}{x}_n)} ,$$

at least if $g^{*}_{nk} \neq 0$. Inserting this into (2.12) gives us (2.11). If either $f_n(\underset{\sim}{x}_n)$ or $g^{*}_{nk} \neq 0$ we must have either g^{**}_{n+k} or $h_{n+k} = 0$ and the factorisation holds trivially. □

Remark It follows from the proof of this and that of (2.9) that we can make this factorisation with the following special choices of f_n and h_{n+k}:

$$f_n(\underset{\sim}{x}_n) = P(\underset{\sim}{x}_n \mid t_n(\underset{\sim}{x}_n))$$

$$h_{n+k}(P, t_{n+k}(\underset{\sim}{x}_{n+k})) = P(t_{n+k}(\underset{\sim}{x}_{n+k})).$$

We can now sharpen our proposition 2.7 to other measures than product measures by using total sufficiency:

2.16 Proposition *If $\underset{\sim}{t}$ is algebraically transitive and t_n is totally sufficient for all* n, *then $\underset{\sim}{t}$ is sufficient and transitive.*

Proof Since total sufficiency for each n implies

$$\underset{\sim}{X}_n \overset{Y_n}{\perp}_P X_{n+1},$$

we use property (0.5) of conditional independence together with the fact that $t_{n+1}(\underset{\sim}{X}_{n+1}) = \phi_{n+1}(Y_n, X_{n+1})$ to get

$$\underset{\sim}{X}_n \perp_P \phi_{n+1}(Y_n, X_{n+1}) = Y_{n+1} \mid Y_n$$

which gives us the transitivity. □

Similarly we have the following strengthening of proposition 2.8.

__2.17 THEOREM__ __The normalised conditional likelihood function is minimal totally sufficient, algebraically transitive and transitive.__

__Proof__ As in proposition 2.8 we just have to prove algebraic transitivity of the normalised conditional likelihood function. The first step is to prove the following relation for the normalisers:

If $\tilde{L}_n(\underset{\sim}{x}_n';\cdot,\cdot) = \tilde{L}_n(\underset{\sim}{x}'';\cdot,\cdot)$

then
$$\frac{\tilde{P}(\underset{\sim}{x}_n')}{\tilde{P}(\underset{\sim}{x}_n', x_{n+1})} = \frac{\tilde{P}(\underset{\sim}{x}_n'')}{\tilde{P}(\underset{\sim}{x}_n'', x_{n+1})} \qquad (2.13)$$

But that follows from the calculation below:

$$\tilde{P}(\underset{\sim}{x}_n', x_{n+1}) = \sup_{\substack{P\in\mathcal{P} \\ k\geqq 2 \\ {}_{n+1}\underset{\sim}{x}_{n+k}}} [P(\underset{\sim}{x}_n', x_{n+1} \mid {}_{n+1}\underset{\sim}{x}_{n+k})]$$

$$= \sup [P(\underset{\sim}{x}_n' \mid {}_n\underset{\sim}{x}_{n+k}) P(x_{n+1} \mid {}_{n+1}\underset{\sim}{x}_{n+k})]$$

$$= \tilde{P}(\underset{\sim}{x}_n') \sup [\tilde{L}_n(\underset{\sim}{x}_n'; P, {}_n\underset{\sim}{x}_{n+k}) P(x_{n+1} \mid {}_{n+1}\underset{\sim}{x}_{n+k})]$$

$$= \frac{\tilde{P}(\underset{\sim}{x}_n')}{\tilde{P}(\underset{\sim}{x}_n'')} \sup \left[P(\underset{\sim}{x}_n'' \mid {}_n\underset{\sim}{x}_{n+k}) P(x_{n+1} \mid {}_{n+1}\underset{\sim}{x}_{n+k})\right]$$

$$= \frac{\tilde{P}(\underset{\sim}{x}_n')}{\tilde{P}(\underset{\sim}{x}_n'')} \tilde{P}(\underset{\sim}{x}_n'', x_{n+1}) .$$

Now we get

$$\tilde{L}_{n+1}(\underset{\sim}{x}_n', x_{n+1}; P, {}_{n+1}\underset{\sim}{x}_{n+k}) =$$

$$\frac{P(\underset{\sim}{x}_n', x_{n+1} \mid {}_{n+1}\underset{\sim}{x}_{n+k})}{\tilde{P}(\underset{\sim}{x}_n', x_{n+1})} =$$

$$\tilde{L}_n(\underset{\sim}{x}_n'; P, {}_n\underset{\sim}{x}_{n+k}) \cdot \frac{\tilde{P}(\underset{\sim}{x}_n')}{\tilde{P}(\underset{\sim}{x}_n', x_{n+1}) P(x_{n+1} \mid {}_{n+1}\underset{\sim}{x}_{n+k})}$$

$$= \tilde{L}_n(\underset{\sim}{x}_n''; P, {}_n\underset{\sim}{x}_{n+k}) \frac{\tilde{P}(\underset{\sim}{x}_n'')}{\tilde{P}(\underset{\sim}{x}_n'', x_{n+1}) P(x_{n+1} \mid {}_{n+1}\underset{\sim}{x}_{n+k})}$$

$$= \tilde{L}_{n+1}(\underset{\sim}{x}_n''; P, {}_{n+1}\underset{\sim}{x}_{n+k}) ,$$

which completes the proof. □

Theorem 2.17 is some sort of a main result of this section. We shall, because of that, mainly center our interest on <u>statistics that are injective functions of the normalised conditional likelihoods</u>, knowing that such functions will have all the nicest possible properties we can think of:

1) minimal totally sufficient
2) algebraically transitive
3) sufficient and transitive.

We shall illustrate the notions just discussed by a few examples. First let us consider the following

<u>2.18 Example</u> Let $X_1,\ldots,X_n,\ldots$ be a homogeneous Markov chain with state space $\{0,1\}$ and transition probabilities

$$P_\theta\{X_{n+1}=1|X_n=0\} = P_\theta\{X_{n+1}=0|X_n=1\} = 1-P_\theta\{X_{n+1}=1|X_n=1\}$$

$$= 1-P_\theta\{X_{n+1}=0|X_n=0\} = \theta/(1+\theta)$$

where $\theta\in\mathbb{R}^+$ is unknown and with (stationary) initial distribution

$$P_\theta\{X_1=1\} = P_\theta\{X_1=0\} = \tfrac{1}{2}.$$

Letting

$$t(x,y) = \begin{cases} 1 & \text{if} \quad x\neq y \\ 0 & \text{if} \quad x=y \end{cases}$$

and

$$t_n(\underset{\sim}{x}_n) = t(x_1,x_2)+\ldots+t(x_{n-1},x_n)$$

we can write the joint probability of $\underset{\sim}{X}_n$ as

$$P_\theta(\underset{\sim}{x}_n) = (1+\theta)^{1-n}\,\theta^{t_n(\underset{\sim}{x}_n)}/2.$$

Clearly $t_n(\underset{\sim}{x}_n)$ is minimal sufficient in the usual sense but since it is not algebraically transitive nor transitive it cannot be minimal totally sufficient.

To find the minimal totally sufficient statistic we calculate the conditional distribution of $\underset{\sim}{X}_n$ given ${}_n\underset{\sim}{X}_{n+k}$ as

$$P_\theta(\underset{\sim}{x}_n|{}_n\underset{\sim}{x}_{n+k}) = P_\theta({}_n\underset{\sim}{x}_{n+k})/P_\theta({}_n\underset{\sim}{x}_{n+k}) = (1+\theta)^{-n}\,\theta^{t_n(\underset{\sim}{x}_n)+t(x_n,x_{n+1})}.$$

From the factorisation theorem (Prop. 2.12) it is clear that the statistic

$$s_n(\underset{\sim}{x}_n) = (t_n(\underset{\sim}{x}_n), x_n)$$

is totally sufficient. It is also algebraically transitive since

$$s_{n+1}(\underset{\sim}{x}_{n+1}) = (t_n(\underset{\sim}{x}_n)+t(x_n,x_{n+1}), x_{n+1})$$

and by Proposition 2.16 therefore transitive. Thus there is a chance that it is minimal totally sufficient. Let us calculate the normalised conditional likelihood function. By standard results we get

$$\tilde{P}(\underset{\sim}{x}_n) = \sup_{{}_n\underset{\sim}{x}_{n+k}} \sup_{\theta} P_\theta(\underset{\sim}{x}_n \mid {}_n\underset{\sim}{x}_{n+k})$$

$$= \begin{cases} \left(\frac{n-t}{n}\right)^n \left(\frac{t}{n-t}\right)^t & \text{if } t \leqq \frac{n-1}{2} \\ \left(\frac{n-t-1}{n}\right)^n \left(\frac{t+1}{n-t-1}\right)^{t+1} & \text{if } t \geqq \frac{n-1}{2} \end{cases}$$

where $t=t_n(\underset{\sim}{x}_n)$. Thus the normalised conditional likelihood function is if $t \leqq \frac{n-1}{2}$:

$$\tilde{L}(\underset{\sim}{x}_n;\theta,{}_n\underset{\sim}{x}_{n+k}) = \frac{(1+\theta)^{-n}\,\theta^{t_n(\underset{\sim}{x}_n)+t(x_n,x_{n+1})}\,n^n}{t_n(\underset{\sim}{x}_n)^{t_n(\underset{\sim}{x}_n)}\,(n-t_n(\underset{\sim}{x}_n))^{n-t_n(\underset{\sim}{x}_n)}}$$

and similarly if $t \geqq \frac{n-1}{2}$. $\tilde{L}(\underset{\sim}{x}_n;\cdot,\cdot)$ is clearly in a one-to-one correspondence with $s_n(\underset{\sim}{x}_n)$.

If we realise that $t_n(\underset{\sim}{x}_n)$ is the number of changes in the random 0-1 sequence it is not too surprising that the information about the unobserved part is contained in $s_n(\underset{\sim}{x}_n)$ but not in $t_n(\underset{\sim}{x}_n)$ since the latter gives no guide to the value of X_{n+1}.

And let us conclude with a brief discussion of an example with a somewhat different flavour.

2.19 Example Let $X_1,\ldots,X_n,\ldots$ be a sequence of 0-1 valued random variables with distribution given as

$$P_{\alpha,\beta}(\underset{\sim}{x}_n) = Be(\alpha+t_n(\underset{\sim}{x}_n),\beta+n-t_n(\underset{\sim}{x}_n))/Be(\alpha,\beta)$$

where Be denotes the Betafunction

$$Be(x,y) = \int_0^1 \theta^{x-1}(1-\theta)^{y-1}d\theta,$$

α and β are unknown and

$$t_n(\underset{\sim}{x}_n) = x_1+\ldots+x_n.$$

Clearly t_n is minimal sufficient. Since it is also both transitive and algebraically transitive it could very well be totally sufficient. In fact the joint distribution of $\underset{\sim}{X}_n$ and ${}_n\underset{\sim}{X}_{n+k}$ is

$$P_{\alpha,\beta}(\underset{\sim}{x}_n,{}_n\underset{\sim}{x}_{n+k}) = \frac{Be(\alpha+t_n(\underset{\sim}{x}_n)+t_k({}_n\underset{\sim}{x}_{n+k}),\beta+n+k-t_n(\underset{\sim}{x}_n)-t_k({}_n\underset{\sim}{x}_{n+k}))}{Be(\alpha,\beta)}$$

and Proposition 2.11 gives that t_n is totally sufficient. Since it is minimal sufficient it must be minimal totally sufficient and equivalent to the normalised conditional likelihood function.

In this example the dependence structure among the random variable was of a type that made no essential difference between total sufficiency and usual sufficiency.

3. Maximal and Extremal Families

We shall now consider a sequence of sample spaces as before, a sequence of statistics $\underset{\sim}{t}$ and we assume this to be sufficient and transitive.

First we shall be interested in classifying the possible distributions of the sequence $Y_n = t_n(X_n)$, $n \in \mathbb{N}$.

As we have seen in proposition 2.2 not all distributions for $Y_n, n \in \mathbb{N}$ can occur since e.g. it is always a Markov process.

Further, if we calculate the reverse transition probabilities of Y_n, $n \in \mathbb{N}$, we see that because t_n is sufficient for all n, these do not depend on $P \in \mathcal{P}$. In fact, if $n \geqq m$

$$P\{Y_m = y_m \mid Y_n = y_n\} =$$

$$\sum_{t_m(\underset{\sim}{x}_m) = y_m} P\{\underset{\sim}{X}_m = \underset{\sim}{x}_m \mid t_n(\underset{\sim}{X}_n) = y_n\} =$$

$$\sum_{\substack{t_m(\underset{\sim}{x}_m) = y_m \\ {}_m\underset{\sim}{x}_n \in {}_m\underset{\sim}{X}_n}} Q_n^{y_n}(\underset{\sim}{x}_n) = \Pi^{t_m}\Pi^{\underset{\sim}{P}_{mn}} Q_{t_n}(y_n)(y_m),$$

where $Q_n^{y_n}(\underset{\sim}{x}_n) = P\{\underset{\sim}{X}_n = \underset{\sim}{x}_n \mid t_n(\underset{\sim}{X}_n) = y_n\}$ is independent of $P \in \mathcal{P}$.

Also we see that the reverse transition probability matrices

$$\{{}_mQ_n(y_m, y_n)\}_{(y_m, y_n) \in \mathcal{Y}_m \times \mathcal{Y}_n}$$

given as

$${}_mQ_n(y_m, y_n) = P\{Y_m = y_m \mid Y_n = y_n\}$$

are exactly the matrices for the mappings ${}_mQ_n$ that occur in the commutative diagrams (2.3). We shall thus use the notation ${}_mQ_n$ both for the mappings and their matrices.

What we now want to characterise is the set of all Markov probabilities on $\underset{\sim}{Y}$ with a specified system of reverse transition probabilities.

We let $\underset{\sim}{Q} = \{{}_mQ_n,\ m \leqq n,\ m,n \in \mathbb{N}\}$ be a system of matrices

$$ {}_mQ_n = \{{}_mQ_n(y_m,y_n)\}_{(y_m,y_n) \in Y_m \times Y_n} $$

satisfying the following set of conditions:

i) $$ {}_mQ_n(y_m,y_n) \geqq 0. \quad \forall (y_m,y_n) \in Y_m \times Y_n $$

ii) $$ \sum_{y_m \in Y_m} {}_mQ_n(y_m,y_n) = 1 \quad \forall y_n \in Y_n $$

(3.1)

iii) $$ {}_mQ_n\ {}_nQ_r = {}_mQ_r \quad \forall\ m \leqq n \leqq r $$

iv) $$ {}_nQ_n = I_{Y_n} \quad \forall\ n \in \mathbb{N} $$

Let $M_{\underset{\sim}{Q}}(\underset{\sim}{Y})$ be the set of all Markov probabilities on $\underset{\sim}{Y}$ such that $\underset{\sim}{Q}$ is the system of reverse transition probabilities, i.e. such that for $P \in M_{\underset{\sim}{Q}}(y)$ and $P\{Y_n = y_n\} > 0$ we have for all $m \leqq n$

$$ P\{Y_m = y_m \mid Y_n = y_n\} = {}_mQ_n(y_m,y_n) $$

for all $y_m \in Y_m$, $y_n \in Y_n$.

$M_{\underset{\sim}{Q}}(\underset{\sim}{Y})$ can be characterised in the following way:

<u>3.1 Proposition</u> <u>There is a one-to-one correspondence between $M_{\underset{\sim}{Q}}(\underset{\sim}{Y})$ and the set of sequences</u> P_n, $n \in \mathbb{N}$, $P_n \in P(Y_n)$ <u>satisfying</u>

$$\forall\, m \leqq n \quad {}_mQ_n P_n = P_m \tag{3.2}$$

<u>the correspondence being given by</u>

$$P_n(y_n) = P\{Y_n = y_n\}.$$

<u>Remark</u> Thus the elements of $M_{\underset{\sim}{Q}}(\underset{\sim}{Y})$ are given by their one-dimensional marginal distributions.

<u>Proof</u> We first prove that the marginal distributions of Y_n must staisfy (3.2) for all $P \in M_{\underset{\sim}{Q}}(\underset{\sim}{Y})$. But we always have for any $P \in \mathbb{P}(\underset{\sim}{Y})$

$$P\{Y_m = y_m\} = \sum_{y_n} P\{Y_m = y_m \mid Y_n = y_n\}\; P\{Y_n = y_n\}$$

such that if we use that for $m \leqq n$ $P\{Y_m = y_m \mid Y_n = y_n\} = {}_mQ_n(y_m, y_n)$ and let $P_n(y_n) = P\{Y_n = y_n\}$, we get

$$P_m(y_m) = \sum_{y_n} {}_mQ_n(y_m, y_n) P_n(y_n),$$

which is (3.2).

Suppose now conversely that we have given a sequence P_n, $n \in \mathbb{N}$ satisfying (3.2). We want to construct a Markov probability on $\underset{\sim}{Y}$ such that P_n are the marginal distributions of Y_n and ${}_mQ_n$ the reverse transition probabilities. This Markov probability is constructed in the usual way, by specifying the forward transition probabilities and the initial distribution. We define for $m < n$

$$
{}_mP_n(y_m,y_n) = \begin{cases} {}_mQ_n(y_m,y_n)\dfrac{P_n(y_n)}{P_m(y_m)} & \text{if } P_m(y_m) > 0 \\ P_n(y_n) & \text{otherwise} \end{cases} \tag{3.3}
$$

and

$$
{}_nP_n(y_n,y_n') = \begin{cases} 1 \text{ if } y_n = y_n' \\ 0 \text{ otherwise} \end{cases} .
$$

We claim that this system $({}_mP_n,\ m \leqq n)$ is a system of forward transition probabilities.

To show this we first note that for $m \leqq n$

$$
P_m(y_m) = 0 \Rightarrow P_n(y_n)\,{}_mQ_n(y_m,y_n) = 0. \tag{3.4}
$$

This follows from (3.2) since

$$
P_m(y_m) = \sum_{y_n} {}_mQ_n(y_m,y_n)P_n(y_n) = 0
$$

and all terms in the sum must be equal to zero.

Clearly ${}_mP_n(y_m,y_n) \geqq 0$ and if $P_m(y_m) = 0, \sum_{y_n} {}_mP_n(y_m,y_n) = \sum_{y_n} P_n(y_n) = 1.$

If $P_m(y_m) > 0$ we have

$$
\sum_{y_n} {}_mP_n(y_m,y_n) = \sum_{y_n} {}_mQ_n(y_m,y_n)\frac{P_n(y_n)}{P_m(y_m)}
$$

$$
= \frac{1}{P_m(y_m)} P_m(y_m) = 1,
$$

where we have used (3.2).

To see that ${}_mP_n$ satisfy the Chapman-Kolmogorov equations we get if $P_m(y_m) = 0$

$$\sum_{y_n} {}_mP_n(y_m,y_n)\,{}_nP_r(y_n,y_r)$$

$$= \sum_{y_n} P_n(y_n)\,{}_nP_r(y_n,y_r)$$

$$= \sum_{P_n(y_n)>0} {}_nQ_r(y_n,y_r)\,P_r(y_r).$$

But if $P_n(y_n) = 0$ we have from (3.4) that ${}_nQ_r(y_n,y_r)P_r(y_r) = 0$ such that we might as well extend the last sum to all y_n and we get

$$\sum_{y_n} {}_mP_n(y_m,y_n)\,{}_nP_r(y_n,y_r) = \sum_{y_n} P_r(y_r)\,{}_nQ_r(y_n,y_r)$$

$$= P_r(y_r) = {}_mP_r(y_m,y_r)$$

by property ii) of (3.1).

If $P_m(y_m) > 0$ we get

$$\sum_{y_n} {}_mP_n(y_m,y_n)\,{}_nP_r(y_n,y_r)$$

$$= \sum_{y_n} {}_mQ_n(y_m,y_n)\frac{P_n(y_n)}{P_m(y_m)}\,{}_nP_r(y_n,y_r)$$

$$= \sum_{P_n(y_n)>0} {}_mQ_n(y_m,y_n)\,{}_nQ_r(y_n,y_r)\frac{P_r(y_r)}{P_m(y_m)}.$$

Using (3.4) again we see that the sum can be extended to all y_n such that

$$\sum_{y_n} {}_mP_n(y_m,y_n)\,{}_nP_r(y_n,y_r)$$

$$= \sum_{y_n} {}_mQ_n(y_m,y_n)\,{}_nQ_r(y_n,y_r)\frac{P_r(y_r)}{P_r(y_m)}$$

$$= {}_mQ_r(y_m,y_r)\frac{P_r(y_r)}{P_m(y_m)} = {}_mP_r(y_m,y_r),$$

where we have used iii) of (3.1).

Since we have now shown that ${}_mP_n$ satisfy all necessary conditions to be forward transition probabilities we can define a Markov probability on $\underset{\sim}{Y}$ with initial distribution P_1 and transition probabilities ${}_mP_n$. We have to show that this probability has marginal distributions P_n and reverse transition probabilities ${}_mQ_n$. By construction $P\{Y_1 = y_1\} = P_1(y_1)$. We have

$$P\{Y_n = y_n\} = \sum_{y_1} P\{Y_n = y_n | Y_1 = y_1\}P\{Y_1 = y_1\}$$

$$= \sum_{y_1} {}_1P_n(y_1,y_n)P_1(y_1) =$$

$$\sum_{P_1(y_1)>0} {}_1Q_n(y_1,y_n)\frac{P_n(y_n)}{P_1(y_1)}\;P_1(y_1)$$

$$= \sum_{y_1} {}_1Q_n(y_1,y_n)P_n(y_n) = P_n(y_n),$$

where we have used (3.4) and ii) of (3.1).

If $P_n(y_n) > 0$ we have

$$P\{Y_m = y_m | Y_n = y_n\} = \frac{P_m(y_m)\;\;{}_mP_n(y_m,y_n)}{P_m(y_n)}$$

If $P_m(y_m) \neq 0$ this is equal to ${}_mQ_n(y_m,y_n)$ by (3.3). If $P_m(y_m) = 0$ we get from (3.4) that ${}_mQ_n(y_m,y_n) = 0$ and thus that

$$P\{Y_m = y_m \mid Y_n = y_n\} = 0 = {}_mQ_n(y_m, y_n).$$

This completes the proof. □

3.2 Corollary $M_{\underset{\sim}{Q}}(\underset{\sim}{Y})$ is convex.

Proof Condition (3.2) is obviously preserved under convex combinations. □

The extreme points of the convex set $M_{\underset{\sim}{Q}}(\underset{\sim}{Y})$ will have special interest to us and they shall be denoted $E_{\underset{\sim}{Q}}(\underset{\sim}{Y})$.

Remark In general $M_{\underset{\sim}{Q}}(\underset{\sim}{Y})$ might be empty and thus also $E_{\underset{\sim}{Q}}(Y)$ empty. This is e.g. the case if $Y_n = Z$ and

$${}_mQ_{n+1}(y_n, y_{n+1}) = \begin{cases} \frac{1}{2} & \text{if } \ |y_n - y_{n+1}| = 1 \\ 0 & \text{otherwise} . \end{cases}$$

For the sake of brevity we shall write M or E instead of $M_{\underset{\sim}{Q}}(\underset{\sim}{Y})$ $E_Q(Y)$ when misunderstanding is impossible.

Having now shown that M is characterised by families of one-dimensional marginal distributions obeying the consistency condition (3.2) we can proceed in analogy with chapter I.

3.3 Proposition If $\underset{\sim}{t}$ is sufficient and transitive $\Pi^{\underset{\sim}{t}}$'s restriction to P is injective and

$$\Pi^{\underset{\sim}{t}}(P) \subseteq M$$

Proof $\Pi^{\underset{\sim}{t}}$ is injective because Π^{t_n} is and any probability on $\underset{\sim}{X}$ is uniquely determined by its marginal distributions of $\underset{\sim}{X}_n$. We have already argued for the inclusion in the beginning of this section. □

As in chapter I we are interested in finding an inverse to $\Pi^{\underset{\sim}{t}}$.

We define for $R \in M$ the measure $Q_{\underset{\sim}{t}}(R)$ on $\underset{\sim}{X}$ by stating that its marginals should be

$$Q_{\underset{\sim}{t}}(R)\{\underset{\sim}{X}_n = \underset{\sim}{x}_n\} = Q_n^{t_n(\underset{\sim}{x}_n)}(\underset{\sim}{x}_n)\ R_n(t_n(\underset{\sim}{x}_n)) \tag{3.5}$$

$$= Q_{t_n} R_n(\underset{\sim}{x}_n)\ .$$

<u>3.4 Proposition</u> (3.5) <u>defines an injective mapping</u> $Q_{\underset{\sim}{t}} : M \to \mathbb{P}(\underset{\sim}{X})$.

<u>Proof</u> We have to verify that the family of marginal distributions given by (3.5) is a consistent family, i.e. that

$$\Pi^{\underset{\sim}{P}_{mn}}(Q_{t_n} R_n) = Q_{t_m} R_m \quad \text{for } m \leqq n\ .$$

But using the commutativity of the diagrams (2.3) and condition (3.2) we get

$$\Pi^{P_{mn}} Q_{t_n}(R_n) = Q_{t_m} {}_m Q_n(R_n) = Q_{t_m}(R_m)\ .$$

That $Q_{\underset{\sim}{t}}$ is injective follows from the fact that all Q_{t_n} are and that a probability on $\underset{\sim}{X}$ is uniquely determined by its consistent family of marginal distributions of $\underset{\sim}{X}_n$. □

<u>3.5 Proposition</u> $\Pi^{\underset{\sim}{t}}$ <u>is the left inverse to</u> $Q_{\underset{\sim}{t}}$ <u>and</u> $Q_{\underset{\sim}{t}}$ <u>is the left inverse to</u> $\Pi^{\underset{\sim}{t}}|P$:

$$\forall R \in M : \Pi^{\underset{\sim}{t}}(Q_{\underset{\sim}{t}}(R)) = R$$

$$\forall P \in P : Q_{\underset{\sim}{t}}(\Pi^{\underset{\sim}{t}}(P)) = P.$$

Proof This follows from the fact that both $\Pi^{\underset{\sim}{t}}$ and $Q_{\underset{\sim}{t}}$ are defined by the mappings Π^{t_n} and Q_{t_n} on the marginals of $\underset{\sim}{X}_n$ and Y_n respectively. Because from propositions 1.3 and 1.4 in §I we have

$$\Pi^{t_n} Q_{t_n}(R_n) = R_n \quad \text{and} \quad Q_{t_n} \Pi^{t_n}(P_n) = P_n$$

We can now define

3.6 Definition The family $\mathcal{P}$ is said to be <u>maximal</u> w.r.t. $\underset{\sim}{t}$ if

$$\Pi^{\underset{\sim}{t}}(\mathcal{P}) = \mathcal{M} \quad \text{or, equivalently} \quad \mathcal{P} = Q_{\underset{\sim}{t}}(\mathcal{M}).$$

3.7 Definition The family $\mathcal{P}$ is said to be <u>extremal</u> w.r.t. $\underset{\sim}{t}$ if

$$\Pi^{t}(\mathcal{P}) = \mathcal{E} \quad \text{or, equivalently} \quad \mathcal{P} = Q_{\underset{\sim}{t}}(\mathcal{E}) \; .$$

If $\mathcal{P}$ is a maximal family we have

3.8 Proposition *If $\mathcal{P}$ is maximal, $\Pi^{\underset{\sim}{t}}|\mathcal{P}$ and $Q_{\underset{\sim}{t}}$ are bijective and each others inverses.*

Proof Immediate from the definition. □

Remark Note that we in the analogous proposition 1.5 of chapter I also could show that for a maximal family the statistic was minimal sufficient. This is not the case here. As the example below shows, it is quite possible that a maximal family consists of only one probability measure P without $\underset{\sim}{t}$ being degenerate.

3.9 Example Let $X_n = \{0,1\}$, and let P be given such that $(X_n, n \in \mathbb{N})$ are independent identically distributed with $P\{X_n = 1\} = 1 - P\{X_n = 0\} = \frac{1}{2}$ and let

$$t_n(x_n) = x_n \qquad t_n : \mathcal{X}_n \to \{0,1\} = \mathcal{Y}_n$$

t_n is clearly sufficient and transitive for the family $\mathcal{P} = \{P\}$.

The reverse transition probabilities are given as

$${}_mQ_n(y_m, y_n) = P\{X_m = y_m \mid Y = y_n\} = \tfrac{1}{2}$$

for all $m, n(y_m, y_n) \in \{0,1\}^2$.

The equation ${}_mQ_nP_n = P_m$ gives thus that

$$P_m(1) = \tfrac{1}{2}P_n(0) + \tfrac{1}{2}P_n(1)$$

$$P_m(0) = \tfrac{1}{2}P_n(0) + \tfrac{1}{2}P_n(1)$$

which clearly implies $P_m(0) = P_m(1) = \tfrac{1}{2}$ for all m. Thus $\{P\}$ is a maximal (and even extremal) family. t is not minimal sufficient and transitive because the trivial statistic is both sufficient and transitive.

Since the mappings $\Pi^{\underset{\sim}{t}}|\mathcal{P}$ and $Q_{\underset{\sim}{t}}$ are bijective and linear, they preserve convexity and map extreme points into extreme points. We thus have

<u>3.10 Proposition</u> <u>A maximal family</u> $\mathcal{P}$ <u>is convex and its extreme points</u> $E(\mathcal{P})$ <u>form an extremal family</u>

$$E(\mathcal{P}) = Q_{\underset{\sim}{t}}(\mathcal{E}).$$

<u>Proof</u> Immediate from the remarks above. □

4. Limit theorems for maximal and extremal families

In chapter I we could characterise maximal and extremal families in terms of the distribution of the sufficient statistic: in a maximal family, the distribution varied completely unrestricted and in extremal families the distribution of the statistic was degenerate.

In the case we are considering in the present chapter this is not so simple (but certainly more interesting). The set of all possible marginal distributions of Y_n, when Y_n is a Markov process with specified reverse transition probabilities, is convex but certainly not equal to $\mathbb{P}(\mathcal{Y}_n)$ in general. Similarly the distribution of Y_n when $P \in \mathcal{E}$ will seldom be degenerate.

We can however, give a similar characterisation if we look at <u>the asymptotic behaviour</u> of the process $(Y_n)_{n \in \mathbb{N}}$ as n tends to infinity. The technique used to describe this asymptotic behaviour is borrowed from the potential theory of transient Markov chains.

To capture the asymptotic behaviour of $(Y_n, n \in \mathbb{N})$ we introduce the following random variables for $n \geqq m$

$$Z_{mn}(y_m) = {}_mQ_n(y_m, Y_n).$$

4.1 Lemma *The sequence $(Z_{mn}(y_m), n \geqq m)$ is for all $P \in \mathcal{P}$ a bounded, reversed martingale w.r.t. $(Y_n, n \geqq m)$*

Proof Clearly $|Z_{mn}(y_m)| \leqq 1$ such that the random variables are bounded. To show that $Z_{mn}(y_m)$, $n \geqq m$ is a reversed martingale w.r.t. Y_n, $n \geqq \mathbb{N}$ we have to show that

$$E_P[Z_{mn}(y_m)\,|\,Y_{n+1},\ldots] = Z_{m,n+1}(y_m)$$

But using the Markov property of Y_n, $n \in \mathbb{N}$ we get

$$E_P[Z_{mn}(y_m)\,|\,Y_{n+1},\ldots] = E_P[Z_{mn}(y_m)\,|\,Y_{n+1}]$$

$$= \sum_{y_n} {}_mQ_n(y_m,y_n)\; P\{Y_n = y_n\,|\,Y_{n+1}\} =$$

$$\sum_{y_n} {}_mQ_n(y_m,y_n)\,{}_nQ_{n+1}(y_n,Y_{n+1}) =$$

$${}_mQ_{n+1}(y_m,Y_{n+1}) = Z_{m,n+1}(y_m)$$

which was to be proved. □

We thus get

4.2 Proposition *The limit as n tends to infinity* $\lim_{n\to\infty} Z_{mn}(y_m) = Z_{m\infty}(y_m)$ *exists almost surely as well as in the mean and*

$$E_P Z_{m\infty}(y_m) = P_m(y_m).$$

Proof That the limit exists a.s. as well as in the mean is a consequence of lemma 4.1 and the reversed martingale convergence theorem. Since we have convergence in the mean we get

$$E_P Z_{m\infty}(y_m) = \lim_{n\to\infty} E_P Z_{mn}(y_m)$$

$$= \lim_{n\to\infty} \sum_{y_n} {}_mQ_n(y_m,y_n)P_n(y_n) =$$

$$\lim_{n\to\infty} P_m(y_m) = P_m(y_m). \quad \square$$

Remark Note that the null-sets $N_{m,y_m} \subset \underset{\sim}{X}$ where convergence does not take place are not more than countably many such that if we let

$$N = \bigcup_m \bigcup_{y_m \in \mathcal{Y}_m} N_{m,y_m}$$

We again get a null-set. This means that the a.s. convergence holds for the $Z_{mn}(y_m)$ simultaneously for all m and y_m.

We thus see that different P's in $\mathcal{P}$ have different limiting distributions for the random variables $Z_{mn}(y_m)$ so that their asymptotic behaviour actually characterises P. In fact $Z_{m\infty}(y_m)$ even have different expectations.

The extreme points of a maximal family $\mathcal{P}$ can be characterised in terms of the behaviour of the random variables $Z_{m\infty}(y_m)$.

To do this we introduce the tail-σ-algebra of the Y_n-process:

$$\underset{\sim}{A}_Y = \bigcap_{n \geqq 1} \sigma\{Y_n, Y_{n+1}, \dots\} = \bigcap_{n \geqq 1} \underset{\sim}{A}^n_Y$$

Note that the random variables $Z_{m\infty}(y_m)$ are all $A_{\underline{Y}}$-measurable since $Z_{mN}(y_m)$ is $\underset{\sim}{A}^n_Y$-measurable for all $N \geqq n$.

To characterise the extreme points we need the following lemma.

4.3 Lemma If $P \in \mathcal{P}$ and $A \in \underset{\sim}{A}_Y$ such that $P(A) > 0$ then $P\{\,|A\}$ is an element of $Q_t(M)$.

Proof We have to show that $P\{Y_n = y_n | A\}$ fits into the equations (3.2). But the Markov property of $\underset{\sim}{\Pi}^t(P)$ implies for $n \geqq m$

$$P\{Y_m = y_m | Y_n = y_n, A\} = P\{Y_m = y_m | Y_n = y_n\}$$

and thus

$$P\{Y_m = y_m | A\} = \sum_{y_n} P\{Y_m = y_m | Y_n = y_n\} P\{Y_n = y_n | A\}$$

$$= \sum_{y_n} {}_mQ_n(y_m, y_n) P\{Y_n = y_n | A\}$$

which was to be proved. □

We are now ready to show the main limit theorem characterising the extreme points of $\mathcal{P}$.

<u>4.4 Theorem</u> <u>Let $\mathcal{P}$ be a maximal family. For $P \in \mathcal{P}$ the following three statements are equivalent.</u>

i) <u>P is an extreme point of</u> $\mathcal{P}$

ii) <u>The tail σ-algebra</u> $\mathcal{A}_{\underset{\sim}{Y}}$ <u>is trivial w.r.t.</u> P.

iii) $Z_{m\infty}(y_m) = P_m(y_m)$ a.s. P.

<u>Proof</u> <u>i) ⇒ ii).</u> Suppose $P \in E(\mathcal{P})$ and $\mathcal{A}_{\underset{\sim}{Y}}$ is not trivial. Then $\mathcal{A}_{\underset{\sim}{Y}}$ contains a set A such that $0 < P(A) < 1$. By lemma 4.3 we have that $P\{\ |A\}$ and $P\{\ |A^c\}$ both are in $\mathcal{P}$ and clearly

$$P\{\cdot\} = P\{A\} \cdot P\{\ |A\} + P\{A^c\} \cdot P\{\ |A^c\}$$

since $P\{\ |A\} \neq P\{\ |A^c\}$ this contradicts the assumption that $P \in E(\mathcal{P})$. Hence $\mathcal{A}_{\underset{\sim}{Y}}$ cannot contain such a set A and is therefore trivial.

<u>ii) ⇒ iii).</u> $Z_{m\infty}(y_m)$ is $\mathcal{A}_{\underset{\sim}{Y}}$-measurable. Thus if $\mathcal{A}_{\underset{\sim}{Y}}$ is trivial, $Z_{m\infty}(y_m)$ is constant a.s. P and equal to its expectation, which by proposition 4.2 is equal to $P_m(y_m)$.

<u>iii) ⇒ ii)</u> If $P = \alpha P^{(1)} + (1-\alpha)P^{(2)}$, we have for all m, $y_m \in \mathcal{Y}_m$ and $t \in \mathbb{R}$ that $P\{Z_{m\infty}(y_m) \leqq t\} = \alpha P^{(1)}\{Z_{m\infty}(y_m) \leqq t\} + (1-\alpha)P^{(2)}$ $\{Z_{m\infty}(y_m) \leqq t\}$. Using now that $Z_{m\infty}(y_m) = P_m(y_m)$ a.s. P, we get

$$1_{[P_m(y_m),\infty[}(t) = \alpha F^{(1)}(t) + (1-\alpha)F^{(2)}(t)$$

This clearly implies that $F^{(1)}(t) = F^{(2)}(t)$ and therefore

$$P_m(y_m) = E_{P(1)} Z_{m\infty}(y_m) = E_{P(2)} Z_{m\infty}(y_m) =$$

$$P_m^{(1)}(y_m) = P_m^{(2)}(y_m).$$

Since all P's are determined from their one-dimensional marginals, we must have $P = P^{(1)} = P^{(2)}$, and hence that P is extreme. □

Occasionally it is practical to introduce the following σ-algebra for $P \in \mathcal{M}$

$$\mathcal{A}_P = \{A : \underset{\sim}{X}_n \perp_P^{Y_n} 1_A, \forall n \in \mathbb{N}\} .$$

We have immediately that

<u>4.5 Proposition</u> $\mathcal{A}_P \supseteq \mathcal{A}_{\underset{\sim}{Y}} \quad \forall P \in \mathcal{M}$.

<u>Proof</u> Since $(t_n)_{n\in\mathbb{N}}$ is transitive, we have

$$\underset{\sim}{X}_n \perp_P^{Y_n} Y_{n+1}, Y_{n+2}, \ldots \quad \forall P \in \mathcal{M},$$

and the result then follows from the fact that

$$\mathcal{A}_{\underset{\sim}{Y}} \subseteq \sigma(Y_{n+1}, Y_{n+2}, \ldots) \ \forall n \in \mathbb{N} . \quad \square$$

We shall call $\mathcal{A}_P$ the <u>extended tail σ-algebra</u>.

But, in fact $\mathcal{A}_P$ is not terribly much greater than $\mathcal{A}_{\underset{\sim}{Y}}$ considered as a measure algebra. We have namely the following straightforward extension of Theorem 4.4.

<u>4.6 Proposition</u> <u>If</u> $P \in \mathcal{M}$, P <u>is an extreme point if and only if</u> $\mathcal{A}_P$ <u>is P-trivial</u>.

<u>Proof</u> If $\mathcal{A}_P$ is trivial, $\mathcal{A}_{\underset{\sim}{Y}}$ is, since $\mathcal{A}_{\underset{\sim}{Y}} \subseteq \mathcal{A}_P$ by Proposition 4.5 and then P is an extreme point by Theorem 4.4.

If there is an $A \in \mathcal{A}_P$ such that $0 < P(A) < 1$, we get by repeating the argument of Lemma 4.3 that $P\{\,|\,A^c\}$ and $P\{\,|\,A^c\}$ both are elements of $\mathcal{M}$ and we thus have the convex combination

$$P = P\{\,|\,A\} \cdot P(A) + P\{\,|\,A^c\}\ P(A^c)$$

which implies that P cannot be extreme. □

Note especially that if we let $\mathcal{A}_{\underset{\sim}{X}}$ be the tail σ-algebra of the X-process,

$$\mathcal{A}_{\underset{\sim}{X}} = \bigcap_{n \in \mathbb{N}} \sigma(X_n, X_{n+1}, \ldots),$$

we have

<u>4.7 Proposition</u> <u>If for all</u> $n \in \mathbb{N}$, t_n <u>is totally sufficient</u>, <u>then</u>

$$\mathcal{A}_{\underset{\sim}{X}} \subseteq \mathcal{A}_P \quad \underline{\text{for all}}\ P \in \mathcal{M}\ .$$

<u>Proof</u> Total sufficiency implies

$$\underset{\sim}{X}_n \perp_P^{Y_n} X_{n+1}, X_{n+2}, \ldots \ \forall n \in \mathbb{N}\ \forall P \in \mathcal{M}\ .$$

Thus $A_{\underset{\sim}{X}} \subseteq A_P$ because

$$A_{\underset{\sim}{X}} \subseteq \sigma(X_{n+1}, X_{n+2}, \ldots) \ \forall n \in \mathbb{N} . \quad \square$$

And as a corollary we get the following 0 - 1 law:

<u>4.8 Corollary</u> <u>If</u> $P \in E$ <u>and</u> t_n <u>is totally sufficient for all n, then</u>

$$A_{\underset{\sim}{X}} \text{ is P-trivial.}$$

<u>Proof</u> $A_{\underset{\sim}{X}} \subseteq A_P$ by Proposition 4.8. If $P \in E$, A_P is P-trivial by Proposition 4.6, thus $A_{\underset{\sim}{X}}$ is P-trivial. $\square$

Note that the converse to the above result in general is false, since the inclusion in Proposition 4.7 can be strict, even as measure algebras.

5. The topology on $(\dot{\bigcup}_n \mathcal{Y}_n) \cup \mathcal{M}$. Boltzmann laws

In this section we shall introduce a topology building together the state space of the statistics and the space of measures $\mathcal{M}$ in the maximal family. The idea to this construction is taken from Martin boundary theory and has as a primary virtue that the limit theorems from the previous section can be given quite an elegant formulation.

First we can consider the space $\mathcal{M}^*$ of all "subconsistent" sequences $(P_n)_{n\in\mathbb{N}}$ of subprobabilities on $\mathcal{Y}_n$, i.e. satisfying

i) $\quad P_n(y_n) \geqq 0$ for all $y_n \in \mathcal{Y}_n$, $n \in \mathbb{N}$

ii) $\quad \sum_{y_n\in\mathcal{Y}_n} P_n(y_n) \leqq 1$ for all $N \in \mathbb{N}$

iii) $\quad P_m \geqq {}_mQ_nP_n$ for all $m < n$.

$\mathcal{M}^*$ is now equipped with the product topology of all the vague topologies for all n, i.e.

$$(P_n^{(\nu)})_{n\in\mathbb{N}} \to (P_n)_{n\in\mathbb{N}} \quad \text{iff for all } n \in \mathbb{N},$$

$$P_n^{(\nu)}(y_n) \to P_n(y_n).$$

Note that proposition 3.1 gives us the possibility of considering $\mathcal{M}$ as a subspace of $\mathcal{M}^*$ by identifying a measure in $\mathcal{M}$ with the sequence of its one-dimensional marginal distributions.

We then consider the space $\mathcal{Y} = \dot{\bigcup}_{n\in\mathbb{N}} \mathcal{Y}_n$ of the disjoint union of the state spaces $\mathcal{Y}_n$ of the sufficient statistic. We shall see that also this space can be identified with a subspace of $\mathcal{M}^*$. Consider namely for $y \in \mathcal{Y}$ the sequence $(K_n^y)_{n \in \mathbb{N}}$ given as

$$K_n^y(y_n) = \begin{cases} {}_nQ_p(y_n, y) & \text{if } y \in \mathcal{Y}_p \text{ and } n \leq p \\ 0 & \text{if } y \in \mathcal{Y}_p \text{ and } n > p. \end{cases}$$

Clearly, $(K_n^y)_{n \in \mathbb{N}} \in \mathcal{M}^*$, a direct consequence of the fact that ${}_mQ_n$ satisfy the Chapman-Kolmogorov equations. Thus we have defined a map

$$i : \mathcal{Y} \cup \mathcal{M} \to \mathcal{M}^*$$

as

$$i(y) = \begin{cases} (K_n^y)_{n \in \mathbb{N}} & \text{for } y \in \mathcal{Y} \\ y\{Y_n = \cdot\}_{n \in \mathbb{N}} & \text{for } y \in \mathcal{M} \end{cases},$$

which is trivially injective and therefore induces a topology on $\mathcal{Y} \cup \mathcal{M}$.

We readily check that the topology induced on $\mathcal{Y}$ is discrete, on $\mathcal{M}$ is the weak topology and the spaces are linked together by the rule that a sequence $(y^{(\nu)})_{\nu \in \mathbb{N}}$ of elements of $\mathcal{Y}$ converges to a $P \in \mathcal{M}$ iff $y^{(\nu)} \in \mathcal{Y}_{N(\nu)}$, $N(\nu) \to \infty$ and

$${}_nQ_{N(\nu)}(y_n, y^{(\nu)}) \to P_n(y_n) .$$

Parts of proposition 4.2 and theorem 4.4 can now be summarised as

5.1 Corollary *For all $P \in \mathcal{M}$, the limit $Y_\infty = \lim_{n \to \infty} Y_n$ exists P-almost surely and we have $P \in \mathcal{E} \Leftrightarrow P = Y_\infty$ P- almost surely.*

Proof The result follows immediately from the remarks on p. 48, the definition of the topology and of the functions Z_m and the statements in the quoted proposition and theorem. □

For later use we note the following

5.2 Lemma $\mathcal{M}^*$ is compact.

Proof $\mathcal{M}^*$ is obviously a subspace of the compact product space of all subprobabilities on $\mathcal{Y}_n$. To show that it is compact, we have to show that the "subconsistency"-condition is closed. But by Fatou's lemma we get

$$\sum_{y_n} {}_mQ_n(y_m,y_n) \lim_{\nu\to\infty} P_n^{(\nu)}(y_n)$$

$$\leqq \lim_{\nu\to\infty} \sum_{y_n} {}_mQ_n(y_m,y_n) P_n^{(\nu)}(y_n)$$

$$\leqq \lim_{\nu\to\infty} P_m^{(\nu)}(y_m) \;,$$

which was to be checked. □

From this lemma we see that we can use this topology to make a quite canonical compactification of the space $\mathcal{Y}$, simply by considering the closure $\overline{\mathcal{Y}}$ of $\mathcal{Y}$ in this topology.

Let $\mathcal{B}^* = \overline{\mathcal{Y}} \setminus \mathcal{Y}$ be the boundary of $\mathcal{Y}$ and consider $\mathcal{B} = \mathcal{B}^* \cap \mathcal{M}$, the set of boundary points with mass unity satisfying the consistency conditions.

We have the following important result

5.3 Proposition $\mathcal{B} \supset \mathcal{E}$, i.e. all extreme points are boundary points of $\mathcal{Y}$.

Proof From theorem 4.4 it follows that for any $P \in \mathcal{E}$ we have $Y_n \to P$ for P-almost all $\underset{\sim}{x} \in \underset{\sim}{X}$. Thus P is a boundary point of $\mathcal{Y}$. □

Proposition 5.3 is sometimes a useful technical tool to actually identify the extreme points. It is occasionally possible to identify all the boundary points and then later relatively easy to pick out those boundary points that are extreme.

Borrowing a terminology from statistical mechanics, we shall call the probability laws in $\mathcal{B}$ for <u>Boltzmann laws</u> (or, more precisely, those in $Q_{\underset{\sim}{t}}(\mathcal{B})$).

As we shall see in later examples we quite frequently have $\mathcal{B} = \mathcal{E}$ although it does not seem to be the case in general.

6. Integral representation of $\mathcal{M}$

To get an integral representation theorem for points in $\mathcal{M}$ in terms of those in $\mathcal{E}$ we run into the difficulty that $\mathcal{M}$ is not a compact set. We therefore take advantage of the fact that $\mathcal{M}^*$ is compact and convex, prove an integral representation theorem for $\mathcal{M}^*$, and then use that $\mathcal{M}$ is a face of $\mathcal{M}^*$ to get an integral representation for $\mathcal{M}$.

Thus we state

6.1 Lemma *$\mathcal{M}^*$ is convex and metrisable and $\mathcal{M}$ is a face in $\mathcal{M}^*$.*

Proof $\mathcal{M}^*$ is clearly convex. The metrisability of $\mathcal{M}^*$ is probably most easily checked by realising that

$$d(P,Q) = \sum_{n\in\mathbb{N}} 2^{-n} \sum_{y_n\in\mathcal{Y}_n} |P_n(y_n) - Q_n(y_n)|$$

is a metric on $\mathcal{M}^*$ inducing the topology in question.

$\mathcal{M}$ is a face of $\mathcal{M}^*$ because, if $P\in\mathcal{M}$ and $P = \lambda P^{(1)} + (1-\lambda)P^{(2)}$, strict inequality in any of the inequalities ${}_mQ_nP_n^{(i)} \leqq P_m^{(i)}$ or $\Sigma P_n^{(i)}(y_n) \leqq 1$ would imply the same strict inequality for P, which cannot be the case for $P\in\mathcal{M}$. Thus $P\in\mathcal{M} \Rightarrow P^{(1)}P^{(2)} \in \mathcal{M}$ and $\mathcal{M}$ is a face. □

It can be seen that this face, even though it is not closed, has the property, that if μ is a probability measure on $\mathcal{M}^*$ with $b(\mu)\in\mathcal{M}$, then $\mu(\mathcal{M}) = 1$.

If E_F and E_K are the extreme points of the face F of K, we note the following

<u>6.2 Lemma</u> <u>If</u> F <u>is a face of</u> K <u>then</u>

$$E_F = F \cap E_K .$$

Proof It is clear that points in $E_K \cap F$ also are in E_F, since points that cannot be obtained by mixing over K certainly not can be mixtures of those in F. Conversely if $x \in E$ and $x = \lambda x_1 + (1 - \lambda) x_2$ then, because F is a face, both x_1 and x_2 are in F. But then we must have $x = x_1 = x_2$ since $x \in E_F$. Thus $x \in E_K$ and the proof is complete. □

We thus have, if we let E^* be the extreme points of M^*:

<u>6.3 Proposition</u> <u>For all</u> $P \in M^*$ <u>there is a probability measure</u> ν <u>on</u> E^* <u>with</u> P <u>as barycenter</u>.

<u>Proof</u> This follows directly from the metrisable version of Choquets theorem, lemma 6.1 and proposition 5.2. □

<u>6.4 Proposition</u> <u>For all</u> $P \in M$ <u>there is a probability measure</u> ν <u>on</u> E <u>with</u> P <u>as barycenter</u>.

<u>Proof</u> By proposition 6.3 there is such a probability ν on E^*. Since $b(\nu) = P \in M$ and M is a face of M^* we must have $\nu(M) = 1$, see the remark before lemma 6.2. By lemma 6.2 we have $E = M \cap E^*$ and thus $\nu(E) = 1$. □

The representation in proposition 6.4 is in fact unique. This can be shown by checking that M is a simplex, but we can also use the fundamental theorem 4.4 to give a direct proof. We are namely able to show that the measure ν can be given a probabilistic interpretation in terms of P and thus that ν is determined by P.

6.5 Proposition For any Borel subset B of E we have

$$\nu(B) = P\{Y_\infty \in B\}.$$

Proof For any $Q \in E$ we have from corollary 5.1 that $Y_\infty = Q$ a.s. and such that

$$Q\{Y_\infty \in B\} = 1_B(Q).$$

If $b(\nu) = P$, we have for any event A that

$$P\{A\} = \int_E Q\{A\}\nu(dQ).$$

Thus,

$$P\{Y_\infty \in B\} = \int_E Q\{Y_\infty \in B\}\nu(dQ)$$

$$= \int_E 1_B(Q)\nu(dQ) = \nu(B). \quad \square$$

6.6 Corollary The representing measure ν in proposition 6.4 is uniquely determined by P:

Proof Proposition 6.5 expresses the ν-measure of any Borel subset of E in terms of P. □

6.7 Corollary For any $P \in M$ we have that $Y_\infty \in E$ a.s. P.

Proof Clearly $1 = \nu(E) = P\{Y_\infty \in E\}$ by proposition 6.5. □

Note that whether or not P is an extreme point, the limit Y_∞ always is and the representing measure ν is simply the distribution of Y_∞.

This fact is analogous to the results in Ch. I in the case with one single experiment. There we had that in a maximal family, the distribution of the sufficient statistic could be anything and in an extremal family, it always had a degenerate distribution. The same thing happens here if we just consider the limiting distribution of the statistic, namely the distribution of Y_∞.

Note that, as a consequence of this, we can never make exact inference on ν from just one sequence of outcomes $(X_n)_{n \in \mathbb{N}}$ in a maximal family.

The only difference in the asymptotic behaviour of Y_n for an extreme point and another $P \in M$ is that Y_∞ takes different values for different sample sequences according to the distribution ν for a non-extreme point whereas $Y_\infty = P$ for almost all sequences when P is an extreme point.

Having in mind that it is quite favourable to be able to make sure inference from complete observation, it is a natural consequence to formulate as a statistical principle always to consider extremal families.

We shall in later examples see what such a principle leads to in a number of interesting cases.

7. Construction of maximal and extremal families

In this section we shall as in I.2 investigate the more unusual aspects of the relation between a statistical model and a sufficient statistic, namely that of constructing or inducing a model corresponding to a given statistic or, rather, a given sequence of statistics. We shall, as in I.2 consider three basic cases:

A. Given the statistics and a system of conditional distributions.

B. Given a statistical model.

C. Given just the sequence of statistics.

We shall first deal with A and then discuss B and C by realising what has to be done to reduce cases B and C to case A.

So let $t_n : \underset{\sim}{X}_n \to Y_n$ be a sequence of surjective statistics and let $\{Q_n^{y_n}, y_n \in Y_n\}$ be a family of distributions $Q_n^{y_n} \in \mathbb{P}(\underset{\sim}{X}_n)$ with $Q_n^{y_n}\{t_n^{-1}(y_n)\} = 1$.

We are looking for probability measures $P \in \mathbb{P}(\underset{\sim}{X})$ such that $Q_n^{y_n}$ are the conditional distributions given t_n, i.e.

$$Q_n^{y_n}(\underset{\sim}{x}_n) = P\{\underset{\sim}{X}_n = \underset{\sim}{x}_n \mid Y_n = y_n\}$$

and further such that $\underset{\sim}{t} = (t_n)_{n \in \mathbb{N}}$ is a <u>transitive</u> sequence, i.e.

$$X_n \perp_P Y_{n+1} \mid Y_n .$$

For any family of such measures, $(t_n)_{n\in\mathbb{N}}$ would be a sufficient and transitive sequence. We are especially interested in extremal families.

For the problem to have a solution at all, we must demand that the system $Q_n^{y_n}$ of conditional distributions satisfies certain regularity conditions.
We have namely earlier shown that <u>transitivity of a sufficient statistic can be expressed by a relation among the</u> $Q_n^{y_n}$-s, see proposition 2.3.

So we demand the following regularity condition to hold for our system $Q_n^{y_n}$:

$$Q_{n+1}^{y_{n+1}}\{\underset{\sim}{X}_n = \underset{\sim}{x}_n\} = Q_n^{t_n(\underset{\sim}{x}_n)}(\underset{\sim}{x}_n)\,Q_{n+1}^{y_{n+1}}\{Y_n = t_n(\underset{\sim}{x}_n)\} \ . \qquad (7.1)$$

But then we form the matrices ${}_mQ_n$ for $m \leqq n$ as

$${}_mQ_n(y_m,y_n) = Q_n^{y_n}\{t_m(\underset{\sim}{X}_m) = y_m\}$$

and know from proposition 2.4 that these must satisfy Chapman-Kolmogorov equations.

We can now determine $M = M_{\underset{\sim}{Q}}(\underset{\sim}{y})$ and $E = E_{\underset{\sim}{Q}}(\underset{\sim}{y})$ as in proposition 3.1.

By proposition 3.5 we have that the families we are looking for are $Q_{\underset{\sim}{t}}(M)$ and $Q_{\underset{\sim}{t}}(E)$, where $Q_{\underset{\sim}{t}}$ is defined as in proposition 3.4.

And, as noted on p. 30, we must remember to be careful, because M and E might be empty or consist of one point only, in which case there is no interesting statistical model with $\underset{\sim}{t}$ as a sufficient and transitive sequence and $Q_n^{y_n}$ as conditionals.

The problem to give useful conditions so as to ensure $\mathcal{E}$ to be non-trivial, i.e. to have more than one point is not easy and is formally equivalent to that of showing the presence of phase transition in statistical mechanics.

It is even a non-trivial problem to characterise those ${}_mQ_n$-systems for which $\mathcal{M}$ is not empty. In one simple case, the answer is affirmative:

<u>7.1 Proposition</u> <u>If $\mathcal{Y}_n$ is finite for all</u> n, $\mathcal{M}$ <u>and</u> $\mathcal{E}$ <u>are non-empty</u>.

<u>Proof</u> As we saw in lemma 5.2, the set $\mathcal{M}^*$ of "subconsistent" subprobabilities is compact. It is also non-empty, the sequences $K^y = (K_n^y)_{n\in\mathbb{N}}$ defined in section 5 are elements of $\mathcal{M}^*$. Now let $y^{(\nu)} \in \mathcal{Y}_\nu$, $\nu = 1,2,\ldots$ be an arbitrary sequence of elements from successive $\mathcal{Y}$-spaces. Consider the corresponding elements $K^{y^{(\nu)}}$ of $\mathcal{M}^*$. Since $\mathcal{M}^*$ is compact $K^{y^{(\nu)}}$ has a convergent subsequence. Let this subsequence have limit K, i.e.

$$K = \lim_{p\to\infty} K^{y^{(\nu_p)}} ,$$

or equivalently

$$K_n(y_n) = \lim_{p\to\infty} {}_nQ_{\nu_p}(y_n, y^{(\nu_p)}) \quad \text{for all } y_n .$$

We shall show that in fact, $K \in \mathcal{M}$. This follows easily by the calculations below, where we use the fact that $\mathcal{Y}_n$ is finite to switch summation and passing to the limit

$${}_mQ_nK_n = {}_mQ_n \lim_{p\to\infty} K_n^{y^{(\nu_p)}}$$

$$= \lim_{p\to\infty} {}_mQ_n \, {}_nQ_{\nu_p} = \lim_{p\to\infty} {}_mQ_{\nu_p} = K_m$$

and

$$\sum_{y_n} K_n(y_n) = \sum_{y_n} \lim_{p\to\infty} {}_nQ_{\nu_p}(y_n, y^{(\nu_p)})$$

$$= \lim_{p\to\infty} \sum_{y_n} {}_nQ_{\nu_p}(y_n, y^{(\nu_p)}) = 1.$$

Thus $K \in M$ and $M \neq \emptyset$.

But according to the integral representation theorem (Proposition 6.4) of section 6, we have

$$K = \int_E Q\nu_k(dQ)$$

for some ν_k, which implies $E \neq \emptyset$. □

So much for case A. In that situation we now know in principle how to construct the maximal and extremal families induced by a sequence of statistics and a consistent system of conditional distributions, just that we have no general effective criterion that ensures such families to be non-empty or non-trivial.

In case B we have given a statistical model, i.e. a family P of probabilities on $\underset{\sim}{X}$. We want to induce a maximal family and an extremal family in such a way that if P itself is maximal or extremal, the induced family is identical to P.

To be able to use the procedure under A, we need a sufficient and transitive sequence. In contrast to our situation in ch. I, we have no guarantee that there is a minimal sequence of statistics that is sufficient and transitive. But theorem 2.17 ensures that we have a minimal totally sufficient sequence that is also sufficient and transitive.

So, we let $(t_n)_{n \in \mathbb{N}}$ be a sequence of minimal totally sufficient statistics and define our system of Q-s as

$$Q_n^{y_n}\{\underset{\sim}{X}_n = \underset{\sim}{x}_n\} = P\{\underset{\sim}{X}_n = \underset{\sim}{x}_n \mid t_n(\underset{\sim}{X}_n) = y_n\}.$$

We have now reduced our problem to case A and therefore we could say that the extremal and maximal families induced by $\mathcal{P}$ are those induced by the above sequence of statistics and Q-system.

We are now met by the difficulty that the extremal and maximal families constructed above might be different from $\mathcal{P}$, even if $\mathcal{P}$ itself was maximal or extremal in advance. This could happen if there was a sequence $\underset{\sim}{t}'$ and a system Q' such that $\underset{\sim}{t}'$ is transitive but not totally sufficient and $\mathcal{P} = Q'_{\underset{\sim}{t}'}(\mathcal{E})$.

In case C we shall assume given a sequence $(t_n)_{n\in\mathbb{N}}$ of statistics, $t_n : \underset{\sim}{\mathcal{X}}_n \to \mathcal{Y}_n$ which is <u>algebraically transitive</u>, i.e.

$$t_{n+1}(\underset{\sim}{x}_{n+1}) = \phi_n(t_n(\underset{\sim}{x}_n), x_{n+1})$$

and we also assume that the set

$$\underset{\sim}{\mathcal{X}}_n^{y_n} = \{\underset{\sim}{x}_n \in \underset{\sim}{\mathcal{X}}_n : t_n(\underset{\sim}{x}_n) = y_n\}$$

is finite and non-empty for all $y_n \in \mathcal{Y}_n$.

We then define our Q-system as the set of <u>microcanonical distributions</u>, i.e.

$$Q_n^{y_n}(\underset{\sim}{x}_n) = 1_{\underset{\sim}{X}_n^{y_n}}(\underset{\sim}{x}_n)/|\underset{\sim}{X}_n^{y_n}|.$$

and we shall then show that our Q-system satisfies condition (7.1) given under A.

If we let

$$_{y_n}X_{n+1}^{y_{n+1}} = \{x_{n+1} \in X_{n+1} : \phi_n(y_n, x_{n+1}) = y_{n+1}\}$$

we get from the algebraic transitivity that

$$1_{\underset{\sim}{X}_{n+1}^{y_{n+1}}}(\underset{\sim}{x}_{n+1}) = \sum_{y_n \in \mathcal{Y}_n} 1_{\underset{\sim}{X}_n^{y_n}}(\underset{\sim}{x}_n) 1_{_{y_n}X_{n+1}^{y_{n+1}}}(x_{n+1})$$

$$= 1_{_{t_n(\underset{\sim}{x}_n)}X_{n+1}^{y_{n+1}}}(x_{n+1}) ,$$

and thus that

$$Q_{n+1}^{y_{n+1}}\{\underset{\sim}{X}_n = \underset{\sim}{x}_n\} = |_{t_n(\underset{\sim}{x}_n)}X_{n+1}^{y_{n+1}}|/|\underset{\sim}{X}_{n+1}^{y_{n+1}}| ,$$

$$Q_n^{t_n(\underset{\sim}{x}_n)}(\underset{\sim}{x}_n) = 1/|\underset{\sim}{X}_n^{t_n(\underset{\sim}{x}_n)}| ,$$

$$Q_{n+1}^{y_{n+1}}\{Y_n = t_n(\underset{\sim}{x}_n)\} = |\underset{\sim}{X}_n^{t_n(\underset{\sim}{x}_n)}| \cdot \frac{|_{t_n(\underset{\sim}{x}_n)}\tilde{X}_{n+1}^{y_{n+1}}|}{|\underset{\sim}{X}_{n+1}^{y_{n+1}}|} ,$$

from which (7.1) follows.

We are then back in case A and define the <u>extremal and maximal families</u> induced by an algebraically transitive sequence of statistics as those induced by the sequence and the corresponding microcanonical distribution.

Also here it might happen that these families are empty or trivial.

In fact, if we induce models from an algebraically transitive system of statistics, t_n will be <u>totally sufficient for all</u> n since for all $P \in \mathcal{M}$, $n,k > 0$:

$$P\{X_1 = x_1, \ldots, X_n = x_{n+k}\} =$$

$$f(P, t_{n+k}(x_1, \ldots, x_{n+k})) = f(P, \psi_{n,k}(t_n(x_n), {}_n\underset{\sim}{x}_{n+k}))$$

which is a factorisation of the type (2.9).

We shall later in this chapter and in chapters III and IV give examples of these constructions. Unfortunately, as also mentioned in the general introduction, the actual identification of the corresponding extremal family is often associated with hard mathematical problems of a technical nature.

8. On the triviality of the tail σ-algebra of a Markov chain

To identify extreme points in any given example, Theorem 4.4 indicates that it is of importance to have conditions for the tail σ-algebra of a non-homogeneous Markov chain to be trivial. Fortunately this problem has been studied rather extensively in the literature, although in most cases, time-homogeneous chains are considered.

On the other hand, the study of homogeneous Markov chains gives quite some insight into non-homogeneous chains, due to the fact, that from any Markov chain $(Y_n, n \in \mathbb{N})$ one can construct a homogeneous chain with "identical" properties, the so-called <u>space-time chain</u>. This construction is made as follows:

Consider our Markov chain $Y_1, Y_2, \ldots$ where Y_n takes values in $\mathcal{Y}_n$ and let

$$P_1(y) = P\{Y_1 = y\} \quad \text{and}$$

$$_nP_{n+k}(y,z) = P\{Y_{n+k} = z \mid Y_n = y\}$$

be initial distribution and transition probabilities of the chain.

Construct now a time-homogeneous chain with state space

$$\mathcal{Y} = \dot{\bigcup_{n \in \mathbb{N}}} \mathcal{Y}_n$$

being the disjoint union of the spaces $\mathcal{Y}_n$, be letting

$$\tilde{Y}_n = i_n(Y_n),$$

where $i_n : \mathcal{Y}_n \to \mathcal{Y}$ is the canonical injection of $\mathcal{Y}_n$ into $\mathcal{Y}$.

$(\tilde{Y}_n, n \in \mathbb{N})$ is obviously a Markov chain with initial distribution

$$\tilde{p}_1(y) = P\{\tilde{Y}_1 = y\} = p_1(y) \cdot 1_{i_1(y_1)}(y)$$

and one-step transition probabilities given as

$$P\{\tilde{Y}_{n+1} = z \mid Y_n = y\} = p(y,z)$$

$$= {}_nP_{n+1}(y,z) \cdot 1_{i_{n+1}(y_{n+1})}(z) \quad \text{if } y \in i_n(y_n)$$

and arbitrary otherwise since $P\{\tilde{Y}_n \in i_n(y_n)\} = 1$ by definition. The tail σ-algebra of Y_n becomes the invariant σ-algebra of $\tilde{Y}_n$.

Using this construction, it is possible to "translate" results for homogeneous chains to non-homogeneous chains, thereby taking advantage of many known results.

Thereby it is important to notice the following facts:

i) A space-time chain is heavily transient, such that only results for <u>transient</u> chains are of interest.

ii) The k-step transition probabilities are given as

$$p^{(k)}(y,z) = {}_nP_{n+k}(y,z) \cdot 1_{i_{n+k}(y_{n+k})}(z) \quad \text{if } y \in i_n(y_n).$$

iii) Many results about triviality of tail σ-algebras give conditions for the tail σ-algebra to be trivial <u>for all</u> initial distributions. Such results cannot be used directly here since we are always interested in a <u>particular</u> initial distribution.

Considerations like the above makes it rather clumsy to use the "translation" procedure in actual cases. It seems more convenient, once and for all to prove the results directly for the non-homogeneous chains, and just use the knowledge that the space-time chain <u>can</u> be constructed as a source of inspiration when guessing which results hold and which do not.

Let now $\mathcal{Y}_n^+(P)$ denote the "effective" state space of Y_n, i.e.

$$\mathcal{Y}_n^+(P) = \{y \in \mathcal{Y}_n : P\{Y_n = y\} > 0\}.$$

We then have the following simple, but quite useful result:

8.1 Proposition $\mathcal{A}_{\underset{\sim}{Y}}$ is P-trivial if and only if for all $n \in \mathbb{N}$, $y', y'' \in \mathcal{Y}_n^+(P)$ and all $A \in \mathcal{A}_{\underset{\sim}{Y}}$:

$$P\{A | Y_n = y'\} = P\{A | Y_n = y''\}. \tag{8.1}$$

Proof (8.1) is equivalent to the statement that 1_A and Y_n are independent. If $\mathcal{A}_{\underset{\sim}{Y}}$ is P-trivial, this is obviously the case. □

Suppose now that 1_A and Y_n are independent for all $n \in \mathbb{N}$. Then, by the Markov property, 1_A is independent of $Y_1, \ldots, Y_n$ for all n. Thus, any random variable, measurable w.r.t.

$$\sigma(Y_1, \ldots, Y_n, \ldots) = \bigcup_n \sigma(Y_1, \ldots, Y_n)$$

are independent of 1_A. In particular, 1_A is independent of itself. Thus

$$P(A \cap A) = P(A)^2 = P(A) \quad ,$$

which implies $P(A) = 0$ or $P(A) = 1$. □

One classical result, cf. Orey (1971), relates the behaviour of bounded so-called space-time harmonic functions to the triviality of the tail σ-algebra.

Call a sequence of functions (real-valued) $f_n, n \in \mathbb{N}$, where f_n is defined on $\mathcal{Y}_n$ P-space-time harmonic if

$${}_nP_{n+k}\, f_{n+k} = f_n \quad \text{a.s.} \;,$$

i.e. if

$$\sum_{z \in \mathcal{Y}_{n+k}} {}_nP_{n+k}(y,z) f_{n+k}(z) = f_n(y) \quad \forall y \in \mathcal{Y}_n^+(P) .$$

We then have

<u>8.2 Proposition</u> $\mathcal{A}_{\underset{\sim}{Y}}$ <u>is P-trivial if and only if all bounded, space-time harmonic functions are constant almost surely.</u>

<u>Proof</u> Suppose first that all bounded, space-time harmonic functions are constant. Then define for $A \in \mathcal{A}_{\underset{\sim}{Y}}, y \in \mathcal{Y}_n^+(P)$

$$f_n^A(y) = P\{A \mid Y_n = y\} .$$

By the Markov-property we have

$$f_n^A(y) = P\{A \mid Y_n = y\} = \sum_z P\{A \mid Y_{n+k} = z\} \cdot P\{Y_{n+k} = z \mid Y_n = y\}$$

$$= \sum_z f_{n+k}^A(z) \, {}_nP_{n+k}(y,z)$$

such that $f_n^A, n \in \mathbb{N}$ is space-time harmonic and also bounded since $|f_n^A| \leqq 1$.

But then it must be constant, and we have

$$f_n^A(y') = f_n^A(y'') , \text{ i.e.}$$

$$P\{A \mid Y_n = y'\} = P\{A \mid Y_n = y''\},$$

which, by Proposition 8.1 implies that $\mathcal{A}_{\underset{\sim}{Y}}$ is trivial.

Conversely, assume $\mathcal{A}_{\underset{\sim}{Y}}$ to be trivial. We first note that f_n is space-time harmonic if and only if

$$F_n = f_n(Y_n), \ n \in \mathbb{N}$$

is a martingale w.r.t. $Y_1, \ldots, Y_n$.

Thus if f_n is bounded and space-time harmonic, F_n converges in the mean and almost surely to a bounded random variable F_∞ where

$$E(F_\infty | Y_n = y) = f_n(y) \text{ a.s.}$$

But F_∞ is tail-measurable and therefore constant almost surely, equal to f, say, whereby $f_n(y) = f$ a.s. which was to be proved. □

Unfortunately the above result is not of much help in any concrete example because of the following reason:

Let f_n be bounded from below by c, say. Then $f_n - c$ is also space-time harmonic, so we might as well assume that $f_n \geqq 0$.

Then, since $F_n = f_n(Y_n)$ is now a non-negative martingale, we can let

$$f = EF_n$$

which is then > 0 unless $f_n \equiv 0$ a.s. .

Let now $(\tilde{P}^f_n)_{n \in \mathbb{N}}$ be defined as

$$\tilde{P}^f_n(y) = P_n(y) \cdot f_n(y)/f \; .$$

We then have $\sum_y \tilde{P}^f_n(y) = 1$ and

$$\begin{aligned}({}_nQ_{n+k}\, \tilde{P}^f_{n+k})(y) &= \frac{1}{f} \sum_z {}_nQ_{n+k}(y,z) P_{n+k}(z) f_{n+k}(z) \\ &= \frac{1}{f} P_n(y) \sum_z {}_nP_{n+k}(y,z) f_{n+k}(z) \\ &= \frac{1}{f} P_n(y) \; f_n(y) = \tilde{P}^f_n(y) \; .\end{aligned}$$

Thus, $(\tilde{P}^f_n)_{n \in \mathbb{N}}$ is an element of M.

On the other hand, let $(\mu_n) \in M$ such that $\mu_n \ll P_n$ for all $n \in \mathbb{N}$.

Then define

$$f_n^{\mu}(y) = \frac{\mu_n(y)}{P_n(y)} \cdot 1_{\mathcal{Y}_n^+(P)}(y)$$

We then conversely have that $(f_n^{\mu})n \in \mathbb{N}$ is space-time harmonic.

There is therefore hardly any difference between finding the space-time harmonic functions and all elements of M and this reduces the usefulness of Proposition 8.2 considerably.

Another classical result relates the behaviour of ${}_nP_{n+k}(y,z)$ for $k \to \infty$ to the triviality of the tail σ-algebra, cf. e.g. Cohn (1974), (1976).

8.3 Proposition $A_{\underset{\sim}{Y}}$ is P-trivial if and only if for all $n \in \mathbb{N}$, $y', y'' \in \mathcal{Y}_n^+(P)$

$$\lim_{k\to\infty} \sum_{z\in\mathcal{Y}_{n+k}} |{}_nP_{n+k}(y',z) - {}_nP_{n+k}(y'',z)| = 0. \tag{8.2}$$

Proof Note that

$$\sum_z |{}_nP_{n+k}(y',z) - {}_nP_{n+k}(y'',z)|$$

$$= \sum_z P_{n+k}(z) \left|\frac{{}_nQ_{n+k}(y',z)}{P_n(y')} - \frac{{}_nQ_{n+k}(y'',z)}{P_n(y'')}\right|$$

$$= E \left|\frac{Z_{n,n+k}(y')}{P_n(y')} - \frac{Z_{n,n+k}(y'')}{P_n(y'')}\right| ,$$

where $Z_{n,n+k}(y) = {}_nQ_{n+k}(y,Y_n)$ are the random variables introduced in section 4.

If P is an extreme point, we get by Theorem 4.4 that $Z_{n,n+k}(y) \to P_n(y)$ almost surely and in the mean, such that the above difference converges to zero.

Conversely, if the difference converges to zero, we have for $A \in \underset{\sim}{A}_Y$

$$|P\{A/Y_n = y'\} - P\{A|Y_n = y''\}|$$

$$= |\sum_z P\{A|Y_{n+k} = z\} \; ({}_nP_{n+k}(y',z) - {}_nP_{n+k}(y'',z))|$$

$$\leqq \sum_z |{}_nP_{n+k}(y',z) - {}_nP_{n+k}(y'',z)| \to 0$$

such that $P\{A|Y_n = y'\} = P\{A|Y_n = y''\}$ and, by Proposition 8.1, P is extreme. □

Again, the condition in Proposition 8.3 is so close to that of Theorem 4.4, that the result is not particularly useful for our purpose. Convergence of $Z_{n,n+k}(y)$ and convergence of their differences are usually equally difficult to prove.

Yet another "classical" result relates the triviality of the tail σ-algebra to a certain mixing condition. We shall abstain from discussing this result here but just mention, that it usually is quite hard to verify that mixing condition.

Recently, coupling techniques have been employed to discuss tail behaviour of Markov chains. It seems that this approach is sufficiently different from those earlier mentioned to be useful. The idea behind these methods, see Griffeath (1975) and Aldous and Pitman (1979), is to make two 'copies' of the original chain Y_n' and Y_n'', say, starting out at different states, each of them evolving according to the transition laws of the original chain, such that they eventually meet and continue together. It is intuitively plausible that this can be done if and only if condition (8.1) is fulfilled, i.e. if and only if the tail σ-algebra of the original chain is trivial.

The "if"-part of the statement is rather hard to prove, but it is of less interest to us. We shall only use the result when we want to show that a given element $P \in M$ in fact is extreme.

Section 9 contains examples of extreme-point models, where the coupling method, although occasionally in a complicated way, works and where the other methods seem to fail in the sense that no one yet has shown how to use them. The success of the coupling method is to a large extent due to the extensive knowledge of the behaviour of random walks, cf. Spitzer (1976), and to results like e.g. the Borel-Cantelli lemma. The drawback of the coupling method, as will be rather clear from our examples, is that there does not seem to be a general technique for constructing such a coupling and that in each case, one has to use a great deal of imagination. In other words, all concrete examples have a tendency to be equipped with much more structure than that of an arbitrary Markov chain, and the coupling method is one way of using this structure more directly than proving one of the previously mentioned almost equivalent results.

To describe the coupling method in somewhat more detail, we introduce the spaces

$$
\begin{aligned}
{}_{n}\tilde{\mathcal{Y}} &= (\mathcal{Y}_n \times \mathcal{Y}_n) \times (\mathcal{Y}_{n+1} \times \mathcal{Y}_{n+1}) \times \cdots \\
&\approx (\mathcal{Y}_n \times \mathcal{Y}_{n+1} \times \cdots) \times (\mathcal{Y}_n \times \mathcal{Y}_{n+1} \times \cdots) \\
&= {}_{n}\mathcal{Y} \times {}_{n}\mathcal{Y}
\end{aligned}
$$

being two "copies" of the spaces ${}_{n}\mathcal{Y}$.

Points in ${}_{n}\tilde{\mathcal{Y}}$ shall be written as

$$
{}_{n}\tilde{y} = ((y_n^{(1)}, y_n^{(2)}), (y_{n+1}^{(1)}, y_{n+1}^{(2)}), \ldots)
$$

or as

$$ {}_n\tilde{\underset{\sim}{y}} = ({}_n\underset{\sim}{y}^{(1)}, {}_n\underset{\sim}{y}^{(2)}) $$

by convenience.

Let now $P \in \mathcal{M}$ and $y', y'' \in \mathcal{Y}_n^+(P)$.

A successful P-<u>coupling</u> of the states y', y'' is a bivariate stochastic process

$$ (Y_0^{(1)}, Y_0^{(2)}), (Y_1^{(1)}, Y_1^{(2)}), \ldots $$

such that $Y_k^{(i)}$ takes values in $\mathcal{Y}_{n+k}$ and such that the joint distribution of this bivariate process $\tilde{P}$ on ${}_n\tilde{\underset{\sim}{\mathcal{Y}}}$ satisfies:

<u>c1</u> : The <u>marginal</u> distributions in $\tilde{P}$ of

$$ (Y_0^{(1)}, Y_1^{(1)}, \ldots) \text{ and } (Y_0^{(2)}, Y_1^{(2)}, \ldots) $$

make $(Y_k^{(i)}, k \in \mathbb{N})$ Markov processes with transition probabilities

$$ \tilde{P}\{Y_{m+k}^{(i)} = z \mid Y_m^{(i)} = y\} = {}_{m+n}P_{m+n+k}(y,z) $$

for all $y \in \mathcal{Y}_{m+n}^+(P)$, $z \in \mathcal{Y}_{m+n+k}$, $m,k \geqq 0$;

<u>c2</u> : The <u>initial</u> distributions satisfy

$$ \tilde{P}\{Y_0^{(1)} = y'\} = 1, \; \tilde{P}\{Y_0^{(2)} = y''\} = 1 $$

<u>c3</u> : There is a random time:

$$ \tau : {}_n\tilde{\underset{\sim}{\mathcal{Y}}} \to \{0,1,\ldots,k,\ldots,\infty\} $$

such that

$$ \tilde{P}\{\tau < \infty\} = 1, \quad \text{and} $$

$$ Y_k^{(1)} = Y_k^{(2)} \quad \text{for all } k > \tau . $$

Note that the joint process $(Y_k^{(1)}, Y_k^{(2)})$ need not be Markov.

We can now prove

8.4 Theorem If $P \in \mathcal{M}$ and and there to all $n \in \mathbb{N}$, $y', y'' \in \mathcal{Y}_n^+(P)$ exists a successful P-coupling, then $\mathcal{A}_{\underset{\sim}{Y}}$ is P-trivial and thus $P \in \mathcal{E}$.

Proof Let $A \in \mathcal{A}_Y$, and $y', y'' \in \mathcal{Y}_n^+(P)$ and suppose we have constructed a successful P-coupling of y' and y''.

Then, by c3, $Y_k^{(1)} = Y_k^{(2)}$ from some point on such that

$$\begin{aligned} &\{(Y_0^{(1)}, Y_1^{(1)}, Y_2^{(1)}, \ldots) \in A\} \\ &= \{(Y_0^{(2)}, Y_1^{(2)}, Y_2^{(2)}, \ldots) \in A\} \text{ a.s. } (\tilde{P}) \end{aligned} \tag{8.3}$$

But, by c1 and c2,

$$\tilde{P}\{(Y_0^{(1)}, Y_1^{(1)}, Y_2^{(1)}, \ldots) \in A\} = P\{A \mid Y_n = y'\}$$

and

$$\tilde{P}\{(Y_0^{(2)}, Y_1^{(2)}, Y_2^{(2)}, \ldots) \in A\} = P\{A \mid Y_n = y''\} .$$

By (8.3) we thus have

$$P\{A \mid Y_n = y'\} = P\{A \mid Y_n = y''\} .$$

Lemma 8.1 now gives that $\mathcal{A}_{\underset{\sim}{Y}}$ is P-trivial. □

8.5 Remark Note, that if we let $y' \sim y''$ if y' and y'' can be coupled, $\sim$ is an equivalence relation.

It can occasionally be easier to construct a coupling that works only with very high probability. We say that a coupling is

<u>(1-ε)-successful</u> if $\tilde{P}\{\tau<\infty\} > 1-\varepsilon$ in the construction just described. We then have

<u>8.6 Theorem</u> <u>If $P \in \mathcal{M}$ and there to all $n \in \mathbb{N}$,</u> $y', y'' \in \mathcal{Y}_n^+(P)$, $\varepsilon>0$ <u>exists a (1-ε)-successful coupling then</u> $A_{\underset{\sim}{Y}}$ <u>is P-trivial and</u> $P \in \mathcal{E}$.

<u>Proof</u>: With the notation as in the proof of the previous theorem we obtain by conditioning upon the event $\{\tau < \infty\}$ that

$$|P\{A|Y_n=y'\}-P\{A|Y_n=y''\}| = |\tilde{P}\{\underset{\sim}{Y}^{(1)} \in A\}-\tilde{P}\{\underset{\sim}{Y}^{(2)} \in A\}|$$

$$= |\tilde{P}\{\underset{\sim}{Y}^{(1)} \in A|\tau=\infty\}\tilde{P}\{\tau=\infty\}-\tilde{P}\{\underset{\sim}{Y}^{(2)} \in A|\tau=\infty\}\tilde{P}\{\tau=\infty\}|$$

$$\leqq 2\tilde{P}\{\tau=\infty\}<2\varepsilon.$$

Since $\varepsilon>0$ was arbitrary, we must have $P\{A|Y_n=y'\} = P\{A|Y_n=y''\}$ and the result follows from Lemma 8.1. □

9. Examples of extremal families

The most important examples of extremal families are of the type where $X_1, \ldots, X_n, \ldots$ is a sequence of independent, identically distributed random variables with distribution unknown and belonging to an exponential family. That is, if the above statement shall be absolutely correct, "exponential family" should mean "generalised exponential family". Because of their special importance, we shall treat these extensively in chapter III, whereas we here shall treat examples of a different type. Unfortunately these tend to be more complicated.

To identify an extremal family, we shall first guess a set of probabilities $P \in M$ that we believe to be extreme points. Then show that this is in fact the case, and here the coupling method discussed in section 8 will be used extensively. Then, finally, we have to show that there are no other extreme points. The latter will usually be done by employing a number of tricks, and it is in general characteristic that each example demands its own technique. This is probably a consequence of the novelty of this line of research. Earlier, many of the examples treated in chapter III were treated separately, each by its own method, whereas now an approach, covering all these examples by one general notion and one technique, is available.

Consider first the following:

<u>9.1 Example</u> Let $\mathcal{X}_n = \{0,1\}$ for all n and let $(\pi_n)_{n \in \mathbb{N}}$ be a fixed sequence of numbers with $\pi_n \in]0,\infty[$.

Assume further that

$$\sum_{n \in \mathbb{N}} \frac{\pi_n}{(1+\pi_n)^2} = \infty \tag{9.1}$$

Let $\mathcal{Y}_n = \{0,1,\ldots,n\}$ and define the sequence of statistics

$$t_n(x_1,\ldots,x_n) = x_1 + \ldots + x_n ,$$

and the system of conditional distributions

$$Q_n^y(x_1,\ldots,x_n) = \frac{\prod_{i=1}^{n} \pi_i^{x_i}}{\gamma_y(\pi_1,\ldots,\pi_n)} 1_{\{y\}}(x_1 + \ldots + x_n) \tag{9.2}$$

where γ_y are the elementary symmetric functions defined as

$$\gamma_s(\lambda_1,\ldots,\lambda_n) = \sum_{\substack{x_1 + \ldots + x_n = s \\ x_i \in \{0,1\}}} \prod_{i=1}^{n} \lambda_i^{x_i} .$$

It is straightforward to verify that this sequence of statistics and corresponding system of conditional distributions satisfy the requirements of section 7. We can thus ask for the extremal family generated by $\underset{\sim}{t}$ and Q.

Consider the family $\{P^\alpha, \alpha \in [0,\infty]\}$ of probabilities on $\underset{\sim}{X} = \{0,1\}^{\mathbb{N}}$ given by

$$P^\alpha(x_1,\ldots,x_n) = \prod_{i=1}^{n} \left(\frac{\alpha\pi_i}{1+\alpha\pi_i}\right)^{x_i} \left(\frac{1}{1+\alpha\pi_i}\right)^{1-x_i} ,$$

i.e. with respect to P^α, the random variables $X_1,\ldots,X_n,\ldots$ are independent and

$$P^\alpha\{X_n = 1\} = 1 - P^\alpha\{X_n = 0\} = \frac{\alpha\pi_n}{1+\alpha\pi_n}$$

such that $\alpha = 0$ and $\alpha = \infty$ correspond to degenerate distributions.

Clearly, $P^\alpha \in M$ for all α, since the conditional distribution of $X_1,\ldots,X_n$ given $X_1+\ldots+X_n=y$, calculated in P^α gives (9.2).

But in fact we have:

9.2 Proposition *The extremal family generated by $\underset{\sim}{t}$ and Q above is the family*

$$\{P^\alpha,\ \alpha\in[0,\infty]\}\ .$$

Proof We first prove that all the P^α-s are extreme, i.e. we shall show that the tail σ-algebra of $Y_n=X_1+\ldots+X_n$, $n\in\mathbb{N}$ is P^α-trivial for all α.

For $\alpha=0$ (or $\alpha=\infty$) this is obvious since then $Y_n=0$ or $Y_n=n$ for all n with P^α-probability one.

We shall now construct a coupling as described in section 8, assuming that $\alpha\in]0,\infty[$.

Then we have $\mathcal{Y}_n^+(P^\alpha)=\mathcal{Y}_n=\{0,1,\ldots,n\}$ for all α.

Let $y',y''\in\mathcal{Y}_n$ for some n and let $(Z_1^{(1)},Z_2^{(1)},\ldots)$ and $(Z_1^{(2)},Z_2^{(2)},\ldots)$ be two independent sequences of independent $\{0,1\}$-valued random variables with

$$\tilde{P}\{Z_k^{(i)}=1\}=\frac{\alpha\pi_{n+k}}{1+\alpha\pi_{n+k}}=1-\tilde{P}\{Z_k^{(i)}=0\}\ .$$

Define further $(Y_0^{(1)},Y_1^{(1)},\ldots)$ and $(Y_0^*,Y_1^*,\ldots)$ as

$$Y_0^{(1)}=y',\quad Y_0^*=y''\quad \text{and}$$

$$Y_k^{(1)} = Y_{k-1}^{(1)} + Z_k^{(1)}, \quad Y_k^* = Y_{k-1}^* + Z_k^{(2)} \quad k \geqq 1 .$$

Define the random time τ as

$$\tau = \inf\{k : Y_k^{(1)} = Y_k^*\}$$

and let

$$Y_k^{(2)} = \begin{cases} Y_k^* & \text{for } k < \tau \\ Y_k^{(1)} & \text{for } k \geqq \tau \end{cases} .$$

It is immediate that if we can show that τ is finite with probability one, we have made a successful P^α-coupling of y' and y''.

But let $D_k = Y_k^{(1)} - Y_k^*$. Then D_k is a Markov chain on the integers with independent increments and

$$P\{D_k - D_{k-1} = 1\} = P\{D_k - D_{k-1} = -1\} = \frac{\alpha\pi_{n+k}}{(1+\alpha\pi_{n+k})^2}$$

and

$$P\{D_k = D_{k-1}\} = \frac{1+\alpha^2\pi^2_{n+k}}{(1+\alpha\pi_{n+k})^2} .$$

D_k, watched only when it jumps, is a simple symmetric random walk which is well known to be recurrent and thus

$$P\{D_k = 0 \text{ infinitely often}\} = 1$$

if and only if

$$P\{D_k \text{ jumps infinitely often}\} = 1 .$$

But

$$\sum_{k=1}^{\infty} P\{D_k - D_{k-1} \neq 0\} = 2 \sum_{k=1}^{\infty} \frac{\alpha\pi_{n+k}}{(1+\alpha\pi_{n+k})^2} ,$$

and using the elementary inequalities

$$\frac{\alpha\pi}{(1+\alpha\pi)^2} \geqq \frac{\pi}{\alpha(1+\pi)^2} \quad \text{if} \quad \alpha \geqq 1$$

$$\frac{\alpha\pi}{(1+\alpha\pi)^2} \geqq \frac{\alpha\pi}{(1+\pi)^2} \quad \text{if} \quad \alpha \leqq 1$$

we get that (9.1) implies

$$\sum_{k=1}^{\infty} P\{D_k - D_{k-1} \neq 0\} = \infty ,$$

and, by the Borel-Cantelli Lemma, that D_k jumps infinitely often. Thus $\tau = \inf\{k : D_k = 0\}$ is finite with probability one and the coupling is successful.

Before we proceed to show that there are in fact no other extreme points, we note that the condition (9.1) is critical, not only to the proof, but also to the result itself. Assume for a while that this is violated, i.e. that

$$\sum_{n=1}^{\infty} \frac{\pi_n}{(1+\pi_n)^2} < \infty \tag{9.3}$$

Then, by the inequalities

$$\frac{\alpha\pi}{(1+\alpha\pi)^2} \leqq \frac{\pi}{\alpha(1+\pi)^2} \quad \text{if} \quad \alpha \leqq 1$$

$$\frac{\alpha\pi}{(1+\alpha\pi)^2} \leqq \frac{\alpha\pi}{(1+\pi)^2} \quad \text{if} \quad \alpha \geqq 1$$

we have that

$$\sum_{n=1}^{\infty} V_\alpha(X_n) = \sum_{n=1}^{\infty} \frac{\alpha\pi_n}{(1+\pi_n)^2} < \infty$$

and thus that

$$Y_N - EY_N = \sum_{n=1}^{N} (X_n - \frac{\alpha\pi_n}{1+\alpha\pi_n})$$

converges a.s. (P^α) to a non-degenerate, tail measurable random variable.

On the other hand, one can allow $\pi_n = 0$ or ∞, without changing the results (nor the proof), just by interpreting the various expressions with π_n in an obvious way.

To show that we have all extreme points, we note the following

<u>9.3 Lemma</u> <u>For all</u> $n \in \mathbb{N}$, $\lambda_1,\ldots,\lambda_n \in \mathbb{R}^+$, $k \leq n$, <u>the functions</u>

$$s \to \frac{\gamma_{s-m}(\lambda_k,\ldots,\lambda_n)}{\gamma_s(\lambda_1,\ldots,\lambda_n)}, \quad s \geq m$$

<u>are non-decreasing in s</u>.

<u>Proof</u> By induction after n. For $n = 1$ the statement is trivial. For $n > 1$ we note that

$$\gamma_s(\lambda_1,\ldots,\lambda_n) = \lambda_n\, \gamma_{s-1}(\lambda_1,\ldots,\lambda_{n-1}) + \gamma_s(\lambda_1,\ldots,\lambda_{n-1}) ,$$

$$\frac{\gamma_{s-m}(\lambda_k,\ldots,\lambda_n)}{\gamma_s(\lambda_1,\ldots,\lambda_n)} = \frac{\lambda_n\gamma_{s-m-1}+\gamma_{s-m}}{\lambda_n\gamma_{s-1}+\gamma_s} = \frac{\lambda_n \frac{\gamma_{s-m-1}}{\gamma_{s-1}} + \frac{\gamma_{s-m}}{\gamma_{s-1}}}{\lambda_n + \frac{\gamma_s}{\gamma_{s-1}}}$$

By the induction hypothesis we thus have that

$$\frac{\gamma_{s-m}(\lambda_k,\ldots,\lambda_n)}{\gamma_s(\lambda_1,\ldots,\lambda_n)} \leq \frac{\lambda_n \frac{\gamma_{s-m}}{\gamma_s} + \frac{\gamma_{s-m+1}}{\gamma_s}}{\lambda_n + \frac{\gamma_{s+1}}{\gamma_s}} = \frac{\lambda_n\gamma_{s-m}+\gamma_{s-m+1}}{\lambda_n\gamma_s + \gamma_{s+1}} = \frac{\gamma_{s+1-m}(\lambda_k,\ldots,\lambda_n)}{\gamma_{s+1}(\lambda_1,\ldots,\lambda_n)}$$

the induction argument then proves the lemma. □

<u>9.4 Corollary</u> <u>The functions</u>

$$P^{\alpha}\{X_n = 1 \mid X_1 = x_1,\ldots,X_{n-1} = x_{n-1},\ Y_N = y\} = \mu_{n,N}(x_1,\ldots,x_{n-1};\ y)$$

<u>are non-decreasing in</u> y, <u>whenever well defined</u>.

<u>Proof</u> The above probability is equal to

$$\frac{\pi_n\ \gamma_{y-\sum_1^{n-1} x_i - 1}(\pi_{n+1},\ldots,\pi_N)}{\gamma_{y-\sum_1^{n-1} x_i}(\pi_n,\pi_{n+1},\ldots,\pi_N)}$$

which is non-decreasing in y by Lemma 9.3. □

Let as usual the Boltzmann laws for this problem be denoted by $\mathcal{B}$ and recall that all extreme laws in $\mathcal{M}$ are Boltzmann laws (Proposition 5.3):

$$\mathcal{E} \subseteq \mathcal{B} \subseteq \mathcal{M}\ .$$

In fact, if $P \in \mathcal{E}$, it follows from Theorem 4.4 that there exists a sequence $y_n' \in Y_n$, such that for all n and y_n

$$\lim_{N\to\infty} P\{Y_n=y_n \mid Y_N=y_N'\} = P\{Y_n=y_n\}$$

and also with $\{Y_n=y_n\}$ replaced by $\{X_1=x_1,\ldots,X_n=x_n\}$.

We have again a lemma:

<u>9.5 Lemma</u> If $P',P'' \in \mathcal{B}$ <u>then we have either</u>

$$P'\{X_n=1 \mid X_1=x_1,\ldots,X_n=x_{n-1}\} \leqq P''\{X_n=1 \mid X_1=x_1,\ldots,X_n=x_{n-1}\}$$

<u>for all</u> $n \in \mathbb{N}$ <u>and</u> $x_1,\ldots,x_{n-1}$ <u>such that they are well defined</u>, <u>or we have everywhere the opposite inequality</u>.

<u>If we have equality everywhere</u>, $P'=P''$.

<u>Proof</u> Choose subsequences (y'_{N_k}) and (y''_{N_ℓ}) such that

$$P'\{x_1,\ldots,x_n\} = \lim_{k\to\infty} P\{X_1=x_1,\ldots,X_n=x_n \mid Y_{N_k}=y'_{N_k}\}$$

and

$$P''(x_1,\ldots,x_n) = \lim_{\ell\to\infty} P\{X_1=x_1,\ldots,X_n=x_n \mid Y_{N_\ell}=y''_{N_\ell}\}$$

Then, clearly

$$P'\{X_n=1 \mid X_1=x_1,\ldots,X_{n-1}=x_{n-1}\} = \lim_{N\to\infty} \mu_{n,N}(x_1,\ldots,x_{n-1};\, y'_N),$$

and similarly for P''.

If the inequality is strict for some n, $x_1,\ldots,x_{n-1}$, we have for this point that

$$\mu_{n,N}(x_1,\ldots,x_{n-1};\, y'_N) < \mu_{n,N}(x_1,\ldots,x_{n-1};\, y''_N)$$

from some point on. But then, by Corollary 9.4, $y'_N < y''_N$ from some point on. Using Corollary 9.4 once more, we see that sine $y'_N < y''_N$

ultimately, the inequality must hold for <u>any</u> $x_1,\ldots,x_{n-1},n$, where the quantities are well defined.

If there is not strict inequality anywhere, we get by an induction argument and the fact that

$$P(x_1,\ldots,x_n) = P(x_n|x_1,\ldots,x_{n-1})\cdot P(x_1,\ldots,x_{n-1})$$

that we must have $P' = P''$. □

Let now $P^* \in \mathcal{B}$ be arbitrary and choose α^* such that

$$P^*\{X_1 = 1\} = \frac{\alpha^*\pi_1}{1+\alpha^*\pi_1} \ .$$

Since the function $\alpha \to \frac{\alpha\pi}{1+\alpha\pi}$ is strictly increasing in α we thus have

$$P^\alpha\{X_1 = 1\} \leqq P^*\{X_1 = 1\} \quad \text{if } \alpha \leqq \alpha^*$$

and

$$P^\alpha\{X_1 = 1\} \geqq P^*\{X_1 = 1\} \quad \text{if } \alpha \geqq \alpha^* \ .$$

Thus, by Lemma 9.5

$$\sup_{\alpha \leqq \alpha^*} P^\alpha\{X_n = 1 | X_1 = x_1,\ldots,X_{n-1} = x_{n-1}\}$$

$$\leqq P^*\{X_n = 1 | X_1 = x_1,\ldots,X_{n-1} = x_{n-1}\}$$

$$\leqq \inf_{\alpha \geqq \alpha^*} P^\alpha\{X_n = 1 | X_1 = x_1,\ldots,X_{n-1} = x_{n-1}\} \ .$$

But we obviously have that the right and left sides of this inequality are equal and equal to

$$P^{\alpha^*}\{X_n = 1 | X_1 = x_1, \ldots, X_n = x_{n-1}\} = \frac{\alpha^* \pi_n}{1+\alpha^* \pi_n} \quad .$$

Thus $P^* = P^{\alpha^*}$ and we have shown that all elements in $\mathcal{B}$ are of the form P^{α} for some $\alpha \in [0,\infty]$.

Since $\mathcal{B} \supseteq \mathcal{E}$ we have now shown

$$\{P^{\alpha}, \alpha \in [0,\infty]\} \subseteq \mathcal{E} \subseteq \mathcal{B} \subseteq \{P^{\alpha}, \alpha \in [0,\infty]\} \tag{9.4}$$

and the proof of Proposition 9.2 is finally completed. □

<u>9.6 Corollary</u> <u>In example</u> 9.1, $\mathcal{B} = \mathcal{E}$.

<u>Proof</u> (9.4). □

<u>9.7 Example</u> (<u>The conditional Rasch model</u>).

One very important example of an extremal family is the family of conditional Rasch models for item analysis, cf. Rasch (1960).

To describe the usual context and notation, consider m problems (items), m fixed, each being answered by n persons, where n is usually large compared to m. The answer can either be 1 or 0 according as the person is solving the problem or not.

Let $X_{\nu i}$, $\nu = 1,\ldots,m$; $i = 1,\ldots,n$ denote the answer of the i'th person to the ν'th question.

A model usually adopted for this experiment is that the $X_{\nu i}$ are independent and

$$P\{X_{\nu i} = 1\} = 1 - P\{X_{\nu i} = 0\} = \frac{\alpha_\nu \beta_i}{1+\alpha_\nu \beta_i} \quad .$$

The parameters $\alpha_\nu, \nu = 1,\dots,m$ are then interpreted as the "difficulty" of the questions and $\beta_1,\dots,\beta_n$ as the "ability" of the persons.

The joint distribution of $\{X_{\nu i}\}$ is then given by

$$P\{X_{\nu i} = x_{\nu i},\ \nu = 1,\dots,m\ ;\ i = 1,\dots,n\}$$

$$= \prod_{i=1}^{n} \prod_{\nu=1}^{m} \frac{(\alpha_\nu \beta_i)^{x_{\nu i}}}{(1+\alpha_\nu \beta_i)} = \frac{\prod_{i=1}^{n} \beta_i^{s_i} \prod_{\nu=1}^{m} \alpha_\nu^{r_\nu}}{\prod_{i=1}^{n} \prod_{\nu=1}^{m} (1+\alpha_\nu \beta_i)}, \tag{9.5}$$

where $s_i = \sum_{\nu=1}^{m} x_{\nu i}$, $r_\nu = \sum_{i=1}^{n} x_{\nu i}$.

Since the joint probability only depends on $(s_1,\dots,s_n;\ r_1,\dots,r_m)$, the corresponding set of statistics is clearly seen to be sufficient and the conditional distribution of $\{X_{\nu i}\}$ given $S_1 = s_1,\dots,S_n = s_n$, $R_1 = r_1,\dots,R_m = r_m$, is the microcanonical:

$$P\{\{x_{\nu i}\} \mid s_1,\dots,s_n;\ r_1,\dots,r_m\}$$

$$= \frac{1}{N_{n,m}(s_1,\dots,s_n;\ r_1,\dots,r_m)}\ 1_A(\{x_{i\nu}\}),$$

where A is the set of 0 - 1 matrices with row-sums $r_1,\dots,r_m$ and column-sums $s_1,\dots,s_n$, and $N_{nm}(\cdot;\cdot)$ is the number of elements in A.

If we let n vary and m be fixed, we see that we have a system of algebraically transitive statistics and corresponding microcanonical distributions, and we can ask for the extremal family corresponding to these.

For several reasons, inference about the 'question-parameters' $\alpha_1,\dots,\alpha_m$ is usually performed in the <u>conditional</u> distribution of $\{X_{\nu i}\}$ given $S_1,\dots,S_n$, this distribution being given as

$$P\{X_{\nu i} = x_{\nu i} \mid S_1 = s_1,\dots,S_n = s_n\}$$

$$= \prod_{i=1}^{n} \frac{\prod_{\nu=1}^{m} \alpha_\nu^{x_{\nu i}}}{\gamma_{s_i}(\alpha_1,\dots,\alpha_m)} = \prod_{i=1}^{n} \frac{\prod_{\nu=1}^{m} \alpha_\nu^{r_\nu}}{\gamma_{s_i}(\alpha_1,\dots,\alpha_m)} . \qquad (9.6)$$

We intend to show on the following pages, that the extremal family consists of these conditional distributions plus certain degenerate measures.

To ease our notation and the connection to the terminology used in the general theory of extremal families, we shall give a slightly different representation of the above set-up.

Rather than representing the answers of a given person by a vector of 0'-s and 1'-s, we shall represent the answers of the i-th person as a <u>subset</u> X_n of the set of questions M, namely that subset of questions he answers correctly, and we define thus our spaces as follows:

Let X_n be the set of subsets of M, where M is a finite set with cardinality $|M| = m$, and define the statistics

$$t_n(\underset{\sim}{x}_n) = (|x_1|,\dots,|x_n|;\ r^{(n)}(\underset{\sim}{x}_n))$$

where

$$r^{(n)}(\underset{\sim}{x}_n) = (r_\mu^{(n)}(\underset{\sim}{x}_n))_{\mu\in M}$$

and

$$r_{\mu}^{(n)}(\underset{\sim}{x}_n) = \sum_{i=1}^{n} 1_{x_i}(\mu) \ .$$

As our system of conditional distributions, we just take the micro-canonical distributions just mentioned.

It will occasionally be practical to consider the statistics

$$t_A(\underset{\sim}{x}_n) = (|x_i|,\ i \in A;\ r^A(\underset{\sim}{x}_n))$$

for $A \subseteq \{1,\ldots,n\}$ where analogously

$$r_{\mu}^{A}(\underset{\sim}{x}_n) = \sum_{i \in A} 1_{x_i}(\mu)$$

are the 'row-sums' of the 'sub-matrix' considering only indices in A. Note that $t_n = t_{\{1,\ldots,n\}}$.

We easily deduce that <u>the conditional distribution of</u> $(X_i, i \in A)$ <u>given</u> t_n <u>only depends on</u> t_A.

Secondly, we note that <u>in any extreme point</u> P, <u>the distribution of</u> $|X_i|$ <u>must be degenerate</u>, since

$$|X_1|,\ldots,|X_n|$$

is a function of t_N for all $N \geq n$ and thus T_N-tail measurable.

Thus, for any extreme point P, there is a particular sequence $\underset{\sim}{s} = (s_1, s_2, \ldots)$ such that $|X_i| = s_i$ for all i.

We can thus partition the extreme points E into disjoint classes $E_{\underset{\sim}{s}}$

$$E = \dot{\bigcup_{\underset{\sim}{s}}} E_{\underset{\sim}{s}} \ ,$$

where $\underset{\sim}{s}$ takes values in all sequences of numbers in $\{0,\ldots,n\}$.

$E_{\underset{\sim}{s}}$ is again the extreme points of $M_{\underset{\sim}{s}}$, where $M_{\underset{\sim}{s}}$ is the set of measures in the maximal family, having $|X_i| = s_i$ degenerate.

We can thus solve our problem by, for fixed $\underset{\sim}{s}$, finding the extremal family $E_{\underset{\sim}{s}}$, with only $r^{(n)}$ as the sufficient statistic.

We can quickly solve the degenerate case, where the set

$$A_+ = \{i : s_i \notin \{0,m\}\}$$

is finite. If we let

$$A_0 = \{i : s_i = 0\}, \quad A_m = \{i : s_i = m\}$$

we see that the variables

$$(r_\mu^{(\infty)})_{\mu\in M} = \lim_{n\to\infty} (r_\mu^{(n)}(\underset{\sim}{x}_n) - |A_m \cap \{1,\ldots,n\}|_{\mu\in M}$$

are tail measurable and finite, and must thus be degenerate in any extreme point whereby the extreme points are given by a vector $r = (r_\mu)_{\mu\in M}$ with

$$P_r\{X_i = \emptyset\} = 1 \quad \text{for } i \in A_0$$

$$P_r\{X_i = M\} = 1 \quad \text{for } i \in A_m$$

and

$$P_r(X_i = x_i;\ i \in A_+) = \frac{1}{N_{|A_+|,M}(s_i, i\in A_+;\ r_\mu, \mu\in M)}$$

if $|x_i| = s_i$ and $r_{A_+}(\underset{\sim}{x}_i, i \in A_+) = r$.

In the following we shall thus concentrate on the case with A_+ infinite.

In fact, one can easily see that we might as well assume A_0, A_m to be empty, since we otherwise can modify the results in an obvious way.

We first show, that the measures in the conditional Rasch model (9.6) are extreme points:

9.8 Proposition *If* $A_0 = A_m = \emptyset$ *the measures*

$$\{P^{\alpha}_{\underset{\sim}{s}},\ \alpha = (\alpha_\mu, \mu \in M) \in (\mathbb{R}^+)^M\}$$

belong to the extremal family $E_{\underset{\sim}{s}}$, *where, according to* $P^{\alpha}_{\underset{\sim}{s}}$, $X_1, X_2, \ldots$ *are independent and*

$$P^{\alpha}_{\underset{\sim}{s}}\{X_n = x_n\} = \frac{\prod_{\mu \in x_n} \alpha_\mu}{\gamma_{s_n}(\alpha_\mu, \mu \in M)} \cdot 1_{\{s_n\}}(|x_n|) \ . \tag{9.7}$$

Proof We have to consider the tail behaviour of the Markov chain $R^{(n)} = r^{(n)}(\underset{\sim}{X}_n)$, $n \in \mathbb{N}$. We shall use a coupling argument to show that the tail σ-algebra is trivial.

Let now

$$A_s = \{n: s_n = s\}$$

and note that $A_0 = A_m = \emptyset$ implies the existence of an s with $|A_s| = \infty$ and $s \notin \{0,m\}$.

So, let $r', r'' \in \mathcal{Y}^+_n(P^{\alpha}_{\underset{\sim}{s}}) = r^{(n)}(\underset{\sim}{X}_n)$. We can without loss of generality assume that

$$r'_\mu = r''_\mu \quad \text{for} \quad \mu \notin \{\kappa, \lambda\}$$

$$r'_\kappa = r''_\kappa + 1, \quad r'_\lambda = r''_\lambda - 1,$$

since $r', r'' \in r^{(n)}(\underset{\sim}{X}_n)$ implies

$$\sum_{\mu \in M} r'_\mu = \sum_{\mu \in M} r''_\mu = \sum_{i=1}^{n} s_i$$

and we can get from any r' to r'' satisfying these constraints by successively adding and subtracting 1 on two coordinates. Further, as noted in Remark 8.5, 'couplability' is an equivalence relation.

Define now two independent processes $(Z_1, Z_2, \ldots)$ and $(\tilde{Z}_1, \tilde{Z}_2, \ldots)$ of independent random variables with

$$P\{Z_k = x\} = P\{\tilde{Z}_k = x\} = P^{\alpha}_{\underset{\sim}{s}}\{X_{n+k} = x\} .$$

Define the event B_k as

$$B_k = \bigcup_{y \subseteq M \setminus \{\kappa, \lambda\}} A_k^y ,$$

where

$$A_k^y = \{Z_k = y \cup \{\kappa\}, \tilde{Z}_k = y \cup \{\lambda\}\} \cup \{Z_k = y \cup \{\lambda\}, \tilde{Z}_k = y \cup \{\kappa\}\} \quad .$$

Note that B_k occurs infinitely often with probability one, since if $|A_s| = \infty$

$$\sum_{k=1}^{\infty} P(B_k) \geqq \sum_{n+k \in A_s} P(A_k^y) = \infty$$

and $(B_k,\ k \geqq 1)$ are independent.

Define further

$$Z_k^* = \begin{cases} Z_k & \text{if not } B_k \\ \tilde{Z}_k & \text{if } B_k \end{cases} ,$$

and note that

$$P\{Z_k^* = x\} = P\{Z_k = x\} ,$$

since this is obvious if x is not of the form $x = y \cup \{\kappa\}$ or $x = y \cup \{\lambda\}$ with $y \subseteq M \setminus \{\kappa,\lambda\}$ and if e.g. $x = y \cup \{\kappa\}$,

$$P\{Z_k^* = y \cup \{\kappa\}\} = P\{(Z_k = y \cup \{\kappa\}) \cap B_k^c\}$$

$$+ P\{(\tilde{Z}_k = y \cup \{\kappa\}) \cap B_k\}$$

$$= P\{Z_k = y \cup \{\kappa\}\} \cdot P\{\tilde{Z}_k \neq y \cup \{\lambda\}\}$$

$$+ P\{\tilde{Z}_k = y \cup \{\kappa\}\} \cdot P\{Z_k = y \cup \{\lambda\}\}$$

$$= P\{Z_k = y \cup \{\kappa\}\} ,$$

so that $(Z_1, Z_2, \ldots)$ and $(Z_1^*, Z_2^*, \ldots)$ are identically distributed (although not independent).

Define now

$$R^{(0)} = r' , \quad \tilde{R}^{(0)} = r'' \quad \text{and}$$

$$R_\mu^{(k+1)} = R_\mu^{(k)} + 1_{Z_k}(\mu)$$

$$\tilde{R}_\mu^{(k+1)} = \tilde{R}_\mu^{(k)} + 1_{Z_k^*}(\mu) .$$

Further, define the random variables

$$D_k = R_\kappa^{(k)} - R_\kappa^{(k)}$$

and let

$$\tau = \inf\{k : D_k = 0\} .$$

Then D_k jumps exactly when B_k occurs, since

$$D_{k+1} - D_k = 1_{Z_k}(\kappa) - 1_{Z_k^*}(\kappa)$$

$$\text{and} \quad B_k = \{Z_k \neq Z_k^*\} .$$

But

$$P\{D_{k+1} - D_k = 1\} = P\{D_{k+1} - D_k = -1\} = \tfrac{1}{2}P(B_k) \quad ,$$

such that, as in the previous example, D_k, watched only when it jumps, is a simple, symmetric random walk on the integers whereby τ is finite with probability one.

Defining now

$$\check{R}^{(k)} = \begin{cases} \tilde{R}^{(k)} & \text{if } k \leqq \tau \\ R^{(k)} & \text{if } k > \tau \end{cases} ,$$

it is evident, that the bivariate process

$$(R^{(0)}, \check{R}^{(0)}),\ (R^{(1)}, \check{R}^{(1)}), \ldots$$

is a successful $P^{\alpha}_{\underset{\sim}{s}}$-coupling of r' and r'' , which implies that $P^{\alpha}_{\underset{\sim}{s}}$ is extreme by Theorem 8.4.

This completes the proof of Proposition 9.8. □

Remark

Aldous and Pitman (1979) gave necessary and sufficient conditions for the exchangeable σ-algebra to be trivial when $X_1, X_2, \ldots$ is a sequence of non-identically distributed, independent random variables taking values is a finite state space. (The exchangeable σ-algebra is the σ-algebra of events left unchanged by a permutation of finitely many coordinates).

It is quite obvious, that the tail σ-algebra of $(R^{(n)}, n \in \mathbb{N})$ is contained in the exchangeable, since $R^{(n)}$ is left unchanged by a permutation of coordinates with index less than or equal to n.

It follows easily by the results of Aldous and Pitman that the exchangeable σ-algebra is $P_{\underset{\sim}{s}}^{\alpha}$-trivial if $A_s = \emptyset$ or $|A_s| = \infty$ for all $s \in \{1,\ldots,m-1\}$. Since the tail σ-algebra of $R^{(n)}$ is contained in the exchangeable, it must itself be trivial in such cases, and $P_{\underset{\sim}{s}}^{\alpha}$ therefore extreme.

On the other hand, if $0 < |A_s| < \infty$ for some s, the same paper shows that the exchangeable σ-algebra is not trivial, such that the above result does not follow from those of Aldous and Pitman.

To find all the extreme points we have to use a number of tricks, utilizing some extra symmetries in the above problem. We note first the following

9.9 Lemma *If A and B are two finite subsets of* $\mathbb{N}$ *and* $\phi: A \to B$ *is bijective, such that for all* $i \in A$,

$$s_{\phi(i)} = s_i ,$$

then, for all $P \in \underset{\sim}{M_s}$, *if* $t_A(\underset{\sim}{x}_n) = t_B(\underset{\sim}{x}_n)$ and $|x_i| = |x_{\phi(i)}|$,

$$P\{X_i = x_i,\ i \in A\} = P\{X_j = x_j,\ j \in B\} . \qquad (9.8)$$

Proof Choose N such that $A \cup B \subseteq \{1,\ldots,N\}$. For all microcanonical distributions we clearly have the relation (9.8) since the number of matrices with given row-column sums does not depend on the labeling of the rows and columns.

But, for any $P \in \underset{\sim}{M_s}$, we have

$$P\{X_i = x_i,\ i \in A\} = \sum_r P\{X_i = x_i,\ i \in A | R^{(N)} = r\} P\{R^{(N)} = r\}$$

$$= \sum_r P\{X_j = x_j,\ j \in B \mid R^{(N)} = r\} P\{R^{(N)} = r\}$$

$$= P\{X_j = x_j,\ j \in B\} . \quad \square$$

In particular, we get e.g. that X_n and $X_{n'}$ are <u>identically distributed</u> if $s_n = s_{n'}$.

Let now $s \in \{1,\dots,m-1\}$ be fixed with $|A_s| = \infty$.

Let further $\mathcal{T}_s$ be the tail σ-algebra of $(X_n, n \in A_s)$.

Clearly, $\mathcal{T}_s \subseteq \mathcal{A}_X$, the tail σ-algebra of the entire sequence of X's.

In fact we can show

<u>9.10 Lemma</u> <u>For any extreme point</u> $P \in \mathcal{E}_s$, $\mathcal{T}_s$ is P-<u>trivial</u>.

<u>Proof</u> Since $R^{(n)}$ is algebraically transitive and the conditional distributions are microcanonical, $R^{(n)}$ is totally sufficient in $\underset{\sim}{\mathcal{M}_s}$ for all n and, by Corollary 4.8, $\mathcal{A}_X$ is P-trivial. Since $\mathcal{T}_s \subseteq \mathcal{A}_X$ it must itself be trivial. $\square$

Further we shall use the following lemma, which easily follows from the fact that a conditional expectation is an L_2-projection:

<u>9.11 Lemma</u> <u>If</u> $\mathcal{A} \subseteq \mathcal{B}$ <u>are</u> σ-<u>algebra's and</u> $EY^2 < +\infty$, <u>then</u>

$$E(E(Y|\mathcal{A})^2) = E(E(Y|\mathcal{B})^2) \Rightarrow E(Y|\mathcal{A}) = E(Y|\mathcal{B}) \text{ a.s.}$$

<u>Proof</u> For a random variable X, let $||X||^2 = EX^2$. We then have

$$||E(Y|\mathcal{B})||^2 = ||E(Y|\mathcal{A})||^2 + ||E(Y|\mathcal{A}) - E(Y|\mathcal{B})||^2 ,$$

from which the result follows. $\square$

Note in particular, that if $E(Y|A)$ and $E(Y|B)$ have the same distribution, the lemma applies and we can conclude that they are identical.

Let now $A_s^n = \{i : s_i = s \wedge i > n\}$.

We then have

9.12 Proposition For any $P \in E_s$ and all $n \in \mathbb{N}$,

$$(X_1,\ldots,X_n) \text{ and } (X_i, i \in A_s^n)$$

are independent.

Proof By Lemma 9.9, the joint distribution of

$$(X_1,\ldots,X_n;\ X_i,\ i \in A)$$

and

$$(X_1,\ldots,X_n;\ X_i, i \in B)$$

are identical if $B \cup A \subseteq A_s^n$ and $|A| = |B| < \infty$.

Thus let $A \in A_s^n$, $B \in A_S^N$ with $N > n$ and let $A \uparrow A_s^n$, $B \uparrow A_S^N$ to get that

$$E(\psi(X_1,\ldots,X_n)|\sigma(X_i, i \in A_s^n))$$

and

$$E(\psi(X_1,\ldots,X_n)|\sigma(X_i, i \in A_S^N))$$

have the same distribution for all ψ.

Apply Lemma 9.11 and let $N \to \infty$ to get

$$E(\psi(X_1,\ldots,X_n)|\sigma(X_i, i \in A_s^n)) = E(\psi(X_1,\ldots,X_n)|T_s) \text{ a.s. } P$$

for any $P \in \underset{\sim}{M_s}$. Now, if $P \in \underset{\sim}{E_s}$, T_s is P-trivial and we thus have

$$E(\psi(X_1,\ldots,X_n)\,|\,\sigma(X_i, i \in A_s^n)) = E(\psi(X_1,\ldots,X_n)) \quad ,$$

showing that $X_1,\ldots,X_n$ and X_i, $i \in A_s^n$ are independent. □

<u>9.13 Corollary</u> <u>If</u> $P \in \underset{\sim}{E_s}$ <u>and</u> $|A_s| = \infty$, $(X_n, n \in A_s)$ <u>are independent and identically distributed</u>.

<u>Proof</u> Follows directly from lemmas 9.9 and 9.12. □

We are now close to showing, that the only non-degenerate extreme points are those of the form $P^{\alpha}_{\underset{\sim}{s}}$ described in Proposition 9.8.

Until further notice, we shall assume that

<u>ND</u> $$\exists s : |A_s| = \infty$$

$$0 < P\{\mu \in X_n\} < 1 \quad \text{for all } n \in A_s,\ \mu \in M \ ,$$

and refer to this as <u>the non-degenerate case</u>.

Define now the coefficients for $\lambda, \mu \in M, n \in A_s$:

$$a_{\lambda\mu} = P\{\lambda \in X_n,\ \mu \notin X_n\} \ .$$

We have

<u>9.14 Lemma</u> <u>In the non-degenerate case, if P is an extreme point</u>

$$a_{\lambda\mu} > 0 \text{ for all } \lambda \neq \mu \ .$$

<u>Proof</u> We have

$$1 = a_{\lambda\mu} + a_{\mu\lambda} + P\{\{\lambda,\mu\} \subseteq X_n\} + P\{\{\lambda,\mu\} \subseteq X_n^c\} \ .$$

If $a_{\lambda\mu} = 0$, we get

$$0 < P\{\lambda \in X_n\} = P\{\{\lambda,\mu\} \subseteq X_n\}$$

which implies

$$P\{\{\lambda,\mu\} \subseteq X_n\} > 0 .$$

Similarly,

$$P\{\{\lambda,\mu\} \subseteq X_n^c\} = 1 - P\{\{\lambda,\mu\} \subseteq X_n\} - a_{\mu\lambda} = 1 - P\{\mu \in X_n\} > 0 .$$

This again implies the existence of $a, b \subseteq M \setminus \{\lambda,\mu\}$ with

$$P\{X_n = a \cup \{\lambda,\mu\}\} > 0 \quad \text{and} \quad P\{X_n = b\} > 0 .$$

But then $|b| = |a| + 2$ and there is an element $\kappa \in b \setminus a$. By Corollary 9.13, we then have for $n, n' \in A_s$:

$$P\{X_{n'} = a \cup \{\lambda,\mu\},\ X_n = b\} = P\{X_n = a \cup \{\lambda,\mu\}\}\ P\{X_n = b\} > 0 ,$$

and, by Lemma 9.9,

$$P\{X_n = a \cup \{\lambda,\mu\}\}\ P\{X_n = b\} = P\{X_n = a \cup \{\kappa,\mu\}\}\ P\{X_n = (b \setminus \{\kappa\}) \cup \{\lambda\}\}$$

which implies

$$0 < P\{X_n = (b \setminus \{\kappa\}) \cup \{\lambda\}\} < a_{\lambda\mu} ,$$

and hence gives a contradiction. □

In the non-degenerate case, we can now strengthen Proposition 9.12:

<u>9.15 Proposition</u> <u>For any non-degenerate extremepoint</u> P, $(X_1, X_2, \ldots)$ <u>are independent and</u>

$$P\{X_n = x\} = P\{X_n = (x \setminus \{\lambda\}) \cup \{\mu\}\} \cdot \frac{a_{\lambda\mu}}{a_{\mu\lambda}} , \tag{9.9}$$

<u>whenever</u> $\lambda \in x$, $\mu \notin x$.

<u>Proof</u> Let $n' \in A_s^n$. By Proposition 9.12 we have

$$P\{X_1 = x_1, \ldots, X_{n-1} = x_{n-1};\ X_n = x, X_{n'} = y\}$$

$$= P\{X_1 = x_1, \ldots, X_{n-1} = x_{n-1}, X_n = x\} \cdot P\{X_{n'} = y\}$$

Letting $y = b \cup \{\mu\}$ with $b \subseteq M \setminus \{\lambda, \mu\}$ and using Lemma 9.8, we get that the above quantity is equal to

$$P\{X_1 = x_1, \ldots, X_{n-1} = x_{n-1}, X_n = (x \setminus \{\lambda\}) \cup \{\mu\}\}\ P\{X_{n'} = b \cup \{\lambda\}\}.$$

Summing over all such b and noticing that

$$a_{\lambda\mu} = \sum_{b \subseteq M \setminus \{\lambda, \mu\}} P\{X_{n'} = b \cup \{\lambda\}\},$$

we thus get

$$P\{X_1 = x_1, \ldots, X_{n-1} = x_{n-1}, X_n = x\}$$

$$= P\{X_1 = x_1, \ldots, X_{n-1} = x_{n-1}, X_n = (x \setminus \{\lambda\}) \cup \{\mu\}\} \cdot \frac{a_{\lambda\mu}}{a_{\mu\lambda}}.$$

Since any set x^* of cardinality s^* can be obtained from any other set x of the same cardinality by substitution of elements one by one in a finite number of steps, we get that if $|x^*| = |x|$

$$P\{X_1 = x_1, \ldots, X_{n-1} = x_{n-1}, X_n = x\} > 0 \Leftrightarrow$$

$$P\{X_1 = x_1, \ldots, X_{n-1} = x_{n-1}, X_n = x^*\} > 0$$

and, if this is the case that

$$\frac{P\{X_n = x\ , X_{n-1} = x_{n-1}, \ldots, X_1 = x_1\}}{P\{X_n = x^*, X_{n-1} = x_{n-1}, \ldots, X_1 = x_1\}}$$

does not depend on $x_1,\ldots,x_{n-1}$, which implies that X_n and $(X_{n-1},\ldots,X_1)$ are independent, whereby the Proposition is proved.□

9.16 Corollary *If* P *is a non-degenerate extreme point.*

$$P\{X_n = x\} > 0 \Leftrightarrow |x| = s_n \ .$$

Proof Immediate from Proposition 9.15. □

The final step before showing that all non-degenerate extreme points are given by (9.7) is to introduce the quantities

$$g_{\lambda\mu} = \begin{cases} \dfrac{a_{\lambda\mu}}{a_{\mu\lambda}} & \text{if } \mu \neq \lambda \\ 1 & \text{if } \mu = \lambda \end{cases} \tag{9.10}$$

and show that these are of an especially simple form. First we show

9.17 Lemma The following relations hold

$$g_{\lambda\mu} = \frac{1}{g_{\mu\lambda}} \tag{9.11}$$

$$g_{\lambda\mu} = g_{\lambda\kappa}\, g_{\kappa\mu} \quad \forall \lambda,\kappa,\mu \in M \tag{9.12}$$

Proof (9.11) is obvious from (9.10).

(9.12) is obvious if either $\lambda = \mu$, $\lambda = \kappa$ or $\kappa = \mu$. So assume that κ,λ,μ are all different.

If $s = 1$ we have

$$a_{\lambda\mu} = P\{\lambda \in X_n,\ \mu \notin X_n\} = P\{\lambda \in X_n\}$$

such that, clearly

$$g_{\lambda\mu} = \frac{P\{\lambda \in X_n\}}{P\{\mu \in X_n\}}$$

which immediately gives (9.12).

If $s \geqq 2$ we have $m \geqq 3$ and there exists, by Corollary 9.16 a set $a \subseteq M \setminus \{\lambda, \kappa, \mu\}$ such that $|a| = s-2$ and

$$P\{X_n = a \cup \{\lambda, \mu\}\} > 0 \quad \text{for } n \in A_s .$$

But then, by Proposition 9.15,

$$P\{X_n = a \cup \{\lambda,\mu\}\} = P\{X_n = a \cup \{\kappa,\mu\}\} \cdot g_{\lambda\kappa}$$

$$= P\{X_n = a \cup \{\kappa,\lambda\}\} \cdot g_{\lambda\kappa}\, g_{\mu\lambda}$$

$$= P\{X_n = a \cup \{\lambda,\mu\}\}\, g_{\lambda\kappa}\, g_{\mu\lambda}\, g_{\kappa\mu} ,$$

i.e. that

$$1 = g_{\mu\lambda}\, g_{\lambda\kappa}\, g_{\kappa\mu} .$$

Dividing by $g_{\mu\lambda}$ and using (9.11) gives (9.12). □

<u>9.18 Corollary</u> <u>There exist numbers</u> $(\alpha_\mu, \mu \in M)$ <u>such that</u>

$$g_{\lambda\mu} = \frac{\alpha_\lambda}{\alpha_\mu} .$$

<u>Proof</u> Choose $\mu_0 \in M$ and let

$$\alpha_\mu = g_{\mu\mu_0}$$

Then, by (9.11) and (9.12)

$$g_{\lambda\mu} = g_{\lambda\mu_0}\, g_{\mu_0\mu} = \frac{g_{\lambda\mu_0}}{g_{\mu\mu_0}} = \frac{\alpha_\lambda}{\alpha_\mu} . \ \square$$

Thus, we can show

9.19 Proposition The non-degenerate extreme points are exactly the measures

$$\{P^{\alpha}_{\underset{\sim}{s}} , \alpha \in (\mathbb{R}^{+})^{M}\}$$

given by (9.8).

Proof They are obviously non-degenerate and extreme by Proposition (9.8).

If, on the other hand, P is non-degenerate and extreme, we can define the numbers $\alpha_\mu, \mu \in M$ from Corollary 9.18 and, by Proposition 9.15 and Corollary 9.16, we then have that, according to P, $(X_1, X_2, \ldots)$ are independently distributed and

$$\frac{P\{X_n = x^*\}}{P\{X_n = x\}} = \frac{\alpha_\lambda}{\alpha_\mu}$$

if $x^* = (x \setminus \{\mu\}) \cup \lambda$, $|x^*| = s_n$.

On the other hand,

$$\frac{P^{\alpha}_{\underset{\sim}{s}}\{X_n = x^*\}}{P^{\alpha}_{\underset{\sim}{s}}\{X_n = x\}} = \frac{\prod_{\nu \in x^*} \alpha_\nu}{\prod_{\nu \in x} \alpha_\nu} = \frac{\alpha_\lambda}{\alpha_\mu}$$

in the same situation. Thus P and $P^{\alpha}_{\underset{\sim}{s}}$ must be identical. □

We shall abstain from the painful task of describing all the degenerate extreme points.

Apart from those, where an entire 'row' is degenerate, i.e. where

$$P\{\lambda \in X_n\} = 0 \quad \text{for all } n \quad \text{or}$$

$$P\{\lambda \in X_n\} = 1 \quad \text{for all } n \quad ,$$

which essentially just reduces the problem to considering a subset of M, we shall mention that e.g. for $m = 4$, the following measure is an extreme point:

$$P\{X_1 = \{1,3,4\}\} = P\{X_1 = \{2,3,4\}\} = \tfrac{1}{2}$$

$$X_2 \equiv X_1^c \cup \{3,4\}$$

for $n \geqq 2$ (X_n) are independent and independent of (X_1, X_2) and further identically distributed as

$$P\{X_3 = \{3\}\} = 1 - P\{X_3 = \{4\}\} = \theta \ .$$

This corresponds to the situation where questions can be arranged in 'blocks', here {1,2} and {3,4}, the relative difficulty of which are infinite and, answering just one question in the 'difficult' group ({1,2}) automatically implies that {3,4} also are answered. Similarly, if not all questions in the 'easy' group ({3,4}) are answered, no questions in a more difficult group are answered either. See also the treatment of Martin-Löf's model in chapter III.

This concludes our treatment of the conditional Rasch model as an extremal family and we now proceed to an example being a simple case of a Poisson regression problem. From that example it should be clear how complicated things can be, even in relative simple set-ups.

9.20 Example (Log-linear Poisson regression)

Let $X_1, X_2, \ldots$ be a sequence of independent Poisson distributed random variables with

$$E_\theta X_n = \theta^{s_n},$$

where $s_1, \ldots, s_n, \ldots$ is a sequence of known real constants and $\theta \geqq 0$ is an unknown parameter.

The joint distribution of $X_1, \ldots, X_n$ is given as

$$P_\theta(\underset{\sim}{x}_n) = \frac{\theta^{\sum_{i=1}^n s_i x_i}}{\prod_{i=1}^n x_i!} e^{-\sum_{i=1}^n \theta^{s_i}},$$

so clearly,

$$t_n(\underset{\sim}{x}_n) = s_1 x_1 + \ldots + s_n x_n$$

is minimal sufficient for θ and totally sufficient for all $n \in \mathbb{N}$.

We could then ask for the extremal family for the above problem and, in particular, ask if the above family is an extreme point family.

It is not surprising, that the solution to this problem depends on the sequence $s_1, \ldots, s_n, \ldots$ but, as we shall see, this dependence is slightly more drastical than one should think by just giving the problem a quick glance.

Suppose for example that $s_n = s$ for all n. Then, in fact, the above model is an extreme point model since it is of the type discussed in chapter III.

If, for example $s_1 = 1$ and $s_n = 2$ for $n \geqq 2$, we see that

$$\sum_{i=1}^{n} s_i X_i \text{ is even for some } n \Leftrightarrow$$

$$\sum_{i=1}^{n} s_i X_i \text{ is even for all } n \Leftrightarrow$$

$$X_1 \text{ is even}$$

Thus the event $\{X_1 \text{ is even}\}$ is tail measurable w.r.t. $\Sigma s_i X_i$ and must therefore be degenerate.

The extreme point family consists then of the measures

$$\{P_\theta^i, i = 1,2;\ \theta \geqq 0\}$$

Where, according to P_θ^i, $(X_n)_{n \in \mathbb{N}}$ are independent, X_n are Poisson with parameter θ^2 for $n \geqq 2$,

$$\left.\begin{aligned} P_\theta^1\{X_1 = 2p\} &= \frac{\theta^{2p}}{(2p!)} \Big/ \sum_{k=0}^{\infty} \frac{\theta^{2k}}{(2k)!} \\ P_\theta^1\{X_1 = 2p+1\} &= 0 \end{aligned}\right\} p = 0,1,\ldots$$

$$\left.\begin{aligned} P^2\{X_1 = 2p+1\} &= \frac{\theta^{2p+1}}{(2p+1)!} \Big/ \sum_{k=0}^{\infty} \frac{\theta^{2k+1}}{(2k+1)!} \\ P_\theta^2\{X_1 = 2p\} &= 0\ . \end{aligned}\right\} p = 0,1,\ldots$$

Similarly, if e.g. $s_n = 2^n$, the events

$$A_n = \{\frac{1}{2^n} \sum_{i=1}^{n} s_i X_i \text{ is even }\}$$

become tail measurable, and the corresponding extremal family gets rather complicated.

We shall now consider <u>the case $s_n = n$</u> in somewhat more detail.

<u>If $\theta = 0$</u>, P_θ is obviously extreme.

<u>If $0 < \theta < 1$</u>, P_θ is <u>not</u> extreme since

$$\sum_{i=1}^{\infty} E|s_i X_i| = \sum_{i=1}^{\infty} i\theta^i < \infty$$

and

$$\sum_{i=1}^{\infty} s_i X_i = Y_\infty$$

is then a tail measurable random variable with a non-degenerate distribution.

We can then get some of the extreme points by considering the measures $\{P^y, y \in \mathbb{N}\}$, where

$$P^y(A) = P_\theta(A \mid Y_\infty = y) \quad .$$

All these P^y-s are extreme, since with P^y-probability one $Y_n = \sum_{i=1}^{n} s_i X_i$ is constant equal to y from some (random) point on.

<u>If $\theta = 1$</u>, P_θ <u>is</u> extreme. To see that this is the case, we construct a coupling of any two points $y', y'' \in \mathcal{Y}_n = \mathbb{N}_0$ as follows:

Define a sequence

$$(Z_1, Z_2, \ldots)$$

of independent, Poisson distributed variables with parameter 1 and let

$$(Z_1^*, Z_2^*, \ldots)$$

be defined as

$$Z_1^* = Z_2, Z_2^* = Z_1, \ldots, Z_{2p+1}^* = Z_{2p+2}, Z_{2p+2}^* = Z_{2p+1}, \ldots \quad .$$

Define now

$$Y'_0 = y', Y^*_0 = y'' \quad \text{and}$$

$$Y'_{k+1} = Y'_k + (n+k+1) Z_{k+1}$$

$$Y^*_{k+1} = Y^*_k + (n+k+1) Z^*_{k+1} .$$

Then, letting $D_k = Y'_{2k} - Y^*_{2k}$, we get

$$\begin{aligned} D_{k+1} &= Y'_{2k+2} - Y^*_{2k+2} \\ &= D_k + (n+2k+1)[Z_{2k+1} - Z^*_{2k+1}] + (n+2k+2)[Z_{2k+2} - Z^*_{2k+2}] \\ &= D_k + Z_{2k+2} - Z_{2k+1} , \end{aligned}$$

since $Z^*_{2k+1} = Z_{2k+2}$ and $Z^*_{2k+2} = Z_{2k+1}$ by construction.

The process $(D_k)_{k \in \mathbb{N}}$ is thus a symmetric random walk on the integers with increments distributed as the difference of two identical Poisson variables. Such a random walk is recurrent, cf. Spitzer (1976), and we thus have, if we let

$$\tau = \inf\{p : p \text{ is even and } D_{\frac{p}{2}} = 0\} ,$$

that $\tau < \infty$ with P_1-probability one.

Defining now

$$Y''_k = \begin{cases} Y^*_k & \text{for } k \leqq \tau \\ Y'_k & \text{for } k > \tau , \end{cases}$$

it is evident that the bivariate process (Y'_0, Y''_0), $(Y'_1, Y''_1), \ldots$ is a successful coupling and P_1 is thus extreme.

The coupling argument provided above does obviously not work for $\theta > 1$. To show that, in fact, <u>if $\theta > 1$ P_θ is extreme</u>, we shall use an $(1-\varepsilon)$-successful coupling as described in theorem 8.6. The idea is due to D. Aldous (personal communication).

We first prove

<u>9.21 Lemma</u> <u>For any</u> $\lambda > 0$ <u>we can construct random variables</u> (X^+, X, X^-) <u>such that each of these are distributed as Poisson with parameter</u> λ <u>and</u>

$$P\{X^+ \neq X+1\} \leqq \frac{1}{\sqrt{\lambda}}$$

$$P\{X^- \neq X-1\} \leqq \frac{1}{\sqrt{\lambda}} \quad .$$

<u>Proof</u> Let ξ be uniform on $]0,1[$ and let F denote the cumulative distribution function of the Poisson distribution with parameter λ,

$$F(t) = \sum_{x \leqq t} P(x) \quad ,$$

where

$$P(x) = \frac{\lambda^x}{x!} e^{-\lambda} \quad .$$

Define $f :]0,1[\to \mathbb{N}_0$ as

$$f(t) = x \quad \text{when } t \in]F(x-1),\ F(x)] \quad ,$$

and let $X = f(\xi)$. X is then obviously Poisson distributed.

A function $f^+ :]0,1[\to \mathbb{N}_0$ is then defined as follows:

$$f^+(t) = x+1 \quad \text{when } t \in]F(x-1), F(x-1) + P(x) \wedge P(x+1)]$$

this defines f^+ everywhere but on the bits

$$\bigcup_{x=0}^{\infty} \left] F(x-1) + P(x) \wedge P(x+1),\ F(x) \right[= R$$

We let now $f^+(t) = 0$ on any measurable subset R_0 of R of Lebesgue measure P(0), and continue to let

$$f^+(t) = 1 \text{ on } R_1 \subseteq R \setminus R_0$$

where R_1 has Lebesgue measure $P(1) - P(0) \wedge P(1)$. In general, let

$$f^+(t) = x+1 \text{ on } R_{x+1} \subseteq R \setminus \left(\bigcup_{i=0}^{x} R_i \right)$$

where R_{x+1} has Lebesgue measure

$$P(x+1) - P(x) \wedge P(x+1) \ .$$

Let now $X^+ = f^+(\xi)$ and we get

$$P\{X^+ = x+1\} = (P(x) \wedge P(x+1)) + P(x+1) - (P(x) \wedge P(x+1))$$

such that X^+ is Poisson distributed. On the other hand, $X^+ = X+1$ everywhere but in R. But the Lebesgue measure of R can be estimated as

$$P\{X^+ \neq X+1\} = \sum_{x=0}^{\infty} (P(x) - P(x) \wedge P(x+1))$$

$$\leqq \sum_{x=0}^{\infty} |P(x) - P(x+1)|$$

$$= \sum_{x=0}^{\infty} \frac{\lambda^{x+1}}{(x+1)!} e^{-\lambda} \left|\frac{x+1-\lambda}{\lambda}\right| \leqq \sum_{x=0}^{\infty} \frac{\lambda^x}{x!} e^{-\lambda} \left|\frac{x-\lambda}{\lambda}\right|$$

$$= \frac{1}{\lambda} E_\lambda |X - \lambda| \leqq \frac{1}{\lambda} \cdot \sqrt{V_\lambda(X)} = \frac{1}{\sqrt{\lambda}} \ .$$

The construction of X^- is analogous and we consider the Lemma as proved. □

And back to the coupling. Let $y', y'' \in Y_n = \mathbb{N}_0$, $y'' = y'+a$, $a>0$, say.

Let then Z_k be independent and Poisson with parameter θ^{n+k}. Choose N so large that

$$N > (\log 4 - 2 \log \varepsilon)/\log \theta - n$$

and define

$$Z_N^* = Z_N^+, \; Z_{N+a}^* = Z_{N+a}^-, \; Z_k^* = Z_k \text{ otherwise,}$$

where Z_k^+, Z_k^- are as in the lemma.

Let further

$$Y_k' = y' + \sum_{i=1}^{k} (n+i) Z_i, \; Y_k'' = y'' + \sum_{i=1}^{k} (n+i) Z_i^*.$$

Then, if we let τ denote the first time that $Y_Y' = Y_k''$ after N+a, we have that they are identical after τ and that

$$P\{\tau<\infty\} \geqq P\{Z_N^*=Z_N+1\} \, P\{Z_{N+a}^*=Z_{N+a}-1\}$$

$$\geqq (1 - \theta^{-\frac{N+n}{2}}) \, (1 - \theta^{-\frac{N+n+a}{2}}) \geqq 1 - \varepsilon.$$

Thus, the coupling is $(1-\varepsilon)$-successful.

We have now shown that <u>the measures</u>

$$\{P_\theta, \; \theta = 0 \quad \text{or} \quad \theta \geqq 1\}$$

<u>as well as</u>

$$\{P^y, \; y \in \mathbb{N}\}$$

<u>where</u> P_θ <u>is given by</u> X_n <u>independent Poisson</u> θ^n <u>and</u>

$$P^y\{\cdot\} = P_{\frac{1}{2}}\{\cdot \mid \sum_{i=1}^{\infty} i X_i = y\} ,$$

<u>all belong to the extremal family</u>.

$P_{\frac{1}{2}}$ could of course have been any P_θ with $\theta < 1$.

We allow ourselves to conjecture that there are no other extreme points. A proof of this result should be obtainable using the techniques developed in Jensen and Johansson (1987), see also chapter IV, section 6.

10. Bibliographical notes

The notion of a sufficient and transitive sequence of statistics (or sub σ-algebras) was introduced and investigated by Bahadur (1954). An algebraically transitive sequence was introduced by Freedman (1962,1963) under the name S-structure. Freedman used this notion to discuss mixtures of Markov chains, and both papers are intimately related to the present work. So is the notions of adequacy introduced by Skibinsky (1967).

Total sufficiency, extremal families of Markov chains were introduced by the author (1972,1973,1974a,1974b) and investigated in some detail. The author was inspired from the work of Martin-Löf (1970) and Martin-Löf (1974) generalised the notion of an extreme family quite drastically.

Most of the above cited work of the author is now parts of sections 2,3 and 4 of the present chapter.

The ideas to all the limit results etc. are taken from Martin Boundary theory, and, as mentioned in section 8, the problem of identifying a given extremal family is almost equivalent to that of identifying a Martin boundary of a suitable Markov chain. In this respect, the author has been inspired by Neveu (1964). Föllmer (1975a,1975b) has used similar ideas in statistical mechanics, and Abrahamse (1970) in the statistical context.

Example 9.1 has been studied by Georgii (1979), Pitman (1978) and Aldous and Pitman (1979) and the proof given here is basically a 'mixture' of Georgii's and Pitman's proof.

J.W. Pitman introduced me to the coupling method and the coupling used to solve the case $\theta = 1$ in example 9.20 is due to him (personal communication). The coupling used to solve $\theta > 1$ in the same example is based on an idea of Aldous (personal communication).

Conditional inference in the Rasch model for item analysis have been studied by Andersen (1973) who showed consistency of the conditional maximum likelihood estimate of the item parameters under the same conditions that ensure the models to be extremal.

The idea used in the proof of Proposition 9.12 is taken from the very interesting paper of Aldous (1981) where he uses the same technique to prove a generalisation of de Finettis theorem which also can be used to give Proposition 9.12.

Most examples in the discrete setting that are discussed in the statistical literature as extreme point models are of the type to be treated in the next section, i.e. corresponding to identical repetitions of experiments with distributions being exponential families in a generalised sense. An exception is the paper by Diaconis and Freedman (1980) where they, in a different 'language' than the present, find the extremal family corresponding to the statistic

$$t_n(x_1,\ldots,x_n) = (n_{xy})_{(x,y)\in \mathcal{X}\times\mathcal{X}}$$

where

$$n_{xy} = \text{the number of } i\text{'s such that } (x_i,x_{i+1}) = (x,y)$$

and show that these are recurrent Markov chains with state space $\mathcal{X}$ and some degenerate processes. See also Zaman (1984) for related results.

The related work of Höglund (1974) shows that in the case where $\mathcal{X}$ is finite and $t : \mathcal{X}\times\mathcal{X} \to Z^d$ is a fixed function, the set of

Boltzmann laws corresponding to the statistics

$$t_n(x_1,\ldots,x_n) = (\sum_{i=1}^{n-1} t(x_i,x_{i+1}),\ x_1,x_n)$$

<u>contains</u> the family of Markov chains with state space X and transition probabilities

$$p_\theta(x,y) = \frac{e^{\langle\theta,t(x,y)\rangle}\, e_\theta(y)}{\phi(\theta)\, e_\theta(x)} \qquad \theta \in \mathbb{R}^d$$

where e_θ is the unique strictly positive eigenvector for the matrix

$$\{e^{\langle\theta,t(x,y)\rangle}\}$$

with eigenvalue $\phi(\theta)$ (being the largest eigenvalue of the matrix).

The full discussion of these "exponential families of Markov chains" has been performed by Ressel (1986) who found all recurrent extreme points in the situation where the state space is a countable set and the statistic t takes values in an Abelian semigroup, in analogy with the models discussed in the next chapter.

Other examples of interest where extreme points have been found typically involve a more complicated repetitive structure and/or a continuous state space and the discussion of these will be postponed to chapter IV.

Chapter III

REPETITIVE STRUCTURES OF POWER TYPE. DISCRETE SAMPLE SPACES

The dominating type of statistic occurring in well known models is _additive_ in an abstract sense, whether this be a counting function such as the _number of times_ certain events occur, the _maximum_ of the observations in a sample or the _frequency table_ (histogram). The fundamental property that these statistics have in common can be expressed by assuming them to take values in an _Abelian semigroup_. In fact one can show that _any statistic which depends symmetrically on the observations and is algebraically transitive, must be of this type_. Thus it seems very worthwhile to devote a large part of the present piece of work to the study of extremal families with such statistics. The models rightly deserve the name of exponential families, where 'exponential' refers to the fact that they are determined by multiplicative homorphisms of the additive semigroup. To us this seems more natural than the classical approach, where vector spaces have been used. We show in section 4 that most results about estimation in exponential families carry over to the more general notion, in a more streamlined fashion, since usual problems with existence of maximum likelihood estimates are avoided. The first section is a compact description of the basic semigroup theory needed.

0. Basic facts about Abelian semigroups

A <u>semigroup</u> is a set S with a composition $S \times S \to S$ which is associative. We shall write this composition as $(s,t) \to st$.

S is <u>Abelian</u> if $st = ts$ for all $s,t \in S$.

Throughout this chapter, "semigroup" means Abelian semigroup.

A <u>unit</u> is S is an element e satisfying $es = s$ for all $s \in S$. If S has a unit, it is unique. If S does not have a unit, such a unit can be adjoined by letting $S^1 = S \cup \{1\}$, where $1s = s$ pr. definition. In the following we shall let $S^1 = S$ if S has a unit or otherwise the above defined semigroup.

An element s of S is <u>idempotent</u> if $s^2 = s$. A semigroup with all elements being idempotent, is a <u>semilattice</u>

A <u>subsemigroup</u> of S is a subset $T \subset S$ closed under composition.

The intersection of subsemigroups is again a subsemigroup such that we can talk about the smallest subsemigroup S_G of S containing a subset $G \subset S$. This is the subsemigroup <u>generated by</u> G. If $S = S_G$, G <u>generates</u> S.

If S and T are semigroups, a map $\phi : S \to T$ is a <u>homomorphism</u> if $\phi(st) = \phi(s)\phi(t)$ for all $s,t \in S$. If $S = T$, ϕ is an <u>endomorphism</u>. If ϕ is <u>bijective</u> it is an <u>isomorphism</u> and so is ϕ^{-1}. S and T are then <u>isomorphic</u>. If $S = T$ and ϕ is an isomorphism, it is an <u>automorphism</u>.

A subset I of S is an <u>ideal</u> if $a \in I$, $b \in S$ implies $ab \in I$. An ideal is <u>completely prime</u> if $ab \in I$ implies either $a \in I$ or $b \in I$. Clearly, an ideal is also a subsemigroup.

A subsemigroup F of S is a <u>filter</u> if for any $a,b \in S$, $ab \in F$ implies $a,b \in F$. $\mathcal{F}(S)$ denotes the filters of S.

A subset F of S is a filter if and only if its complement is a completely prime ideal.

The intersection of filters is a filter and since S is a filter, it makes sense to discuss the <u>smallest filter</u> F_s containing a given element s of S.

A <u>congruence</u> on S is an equivalence relation $\sim$ on S such that $s \sim t$ and $u \sim v$ implies $su \sim tv$.

If $\sim$ is a congruence, the set of $\sim$-equivalence classes form a semi group with the obvious composition, the <u>quotient semigroup</u> $S/\sim$. The canonical map $S \to S/\sim$ that maps an element into its equivalence classes is a homomorphism.

A congruence $\sim$ is <u>idempotent</u> if $s \sim s^2$. Then $S/\sim$ is a <u>semilattice</u> and all equivalence classes are subsemigroups of S. We say that $\sim$ define a <u>semilattice decomposition</u> of S.

The equivalence relation $s \phi t \Leftrightarrow F_s = F_t$ is obviously an idempotent congruence. The ϕ-equivalence classes are called the <u>filter components of</u> S, in the following denoted $\Phi(S)$. A semigroup is cancellative if xs = xt implies $s = t$ for all $x,s,t \in S$.

If a semigroup is cancellative it can be embedded into a group <u>the group of quotients of S</u>, as follows:

Let $S \times S$ denote the direct product of S with itself, being the Cartesian product equipped with the composition

$$(s_1,s_2)\ (t_1,t_2) = (s_1t_1,s_2t_2)$$

and define a congruence q on $S \times S$ as

$$(s_1,s_2)q(t_1,t_2) \Leftrightarrow s_1t_2 = s_2t_1$$

The quotient semigroup $Q(S) = S \times S/q$ is then a group. If $[s,t]$ denotes the q equivalence class of (s,t), we define the map

$$\psi: s \to [s^2,s]$$

ψ is then an isomorphism of S onto $\psi(S)$.

Usually we shall write the equivalence class $[s,t]$ as st^{-1}.

A semigroup is called <u>separative</u> if $st = s^2 = t^2$ implies $s = t$.

A <u>character</u> χ on a semigroup is a multiplicative functional on S into the field of complex numbers

$$\chi(st) = \chi(s)\chi(t)$$

The set of characters of S is denoted S^* and is a semigroup under multiplication.

We say that the characters <u>separate points</u> in S if for any $s,t \in S$, there exists a character $\chi \in S^*$ such that $\chi(s) \neq \chi(t)$.

<u>The characters separate points in</u> S <u>if and only if</u> S <u>is separative</u> (Clifford and Preston (1961), Theorem 5.59).

If χ is a character, then let

$$F^{\chi} = \{s \in S: \ \chi(s) \neq 0\}$$

F^{χ} is obviously a filter, the non-vanishing filter of χ. Similarly we call

$$I^{\chi} = (F^{\chi})^c$$

the <u>vanishing ideal</u> of χ.

If $\chi(a) \neq 0$ then, clearly $F^{\chi} \supseteq F_a$.

An <u>exponential function</u> on S is a real-valued, non-negative character.

The set of exponential functions on S form a semigroup EXP(S) under pointwise multiplication.

We shall say that a semigroup is <u>strongly separative</u> if the exponential functions separate points.

Note that a subset F of S is a filter if and only if there is an exponential function χ such that $F = F^{\chi}$. The indicator $\chi = 1_F$ can be used.

We have

<u>0.1 Theorem</u> S <u>is strongly separative if and only if</u>

a) <u>All filter components are cancellative and</u>

b) <u>All elements of the group of quotients</u> $Q(\Phi)$, $\Phi \in \Phi(S)$ <u>other than the identities have infinite order</u>, i.e. $Q(\Phi)$ <u>is a torsion free group</u>

<u>Proof</u> Assume first that S is strongly separative. Let x,a,b be in the same filter component of S:

$$F_x = F_a = F_b,$$

and assume $a \neq b$, $xa = xb$.

Choose χ to be an exponential function such that $\chi(a) \neq \chi(b)$.

Then if $\chi(a) > 0, F_a \subseteq F^{\chi} = F$ such that $\chi(x) > 0$. We then have

$$\chi(a\,x) = \chi(b\,x) = \chi(a)\chi(x) = \chi(b)\chi(x) \Rightarrow \chi(a) = \chi(b)$$

which is a contradiction.

Let s,t be in the same filter component and assume that st^{-1} has finite order m, $s \neq t$.

Then $(st^{-1})^m = e$ such that

$$s^m = t^m .$$

Then, for any exponential function

$$\chi(s)^m = \chi(t)^m \to \chi(s) = \chi(t)$$

contradicting the assumption of strong separativity.

Assume now that a) and b) are satisfied. Then let $s,t \in S$ with $s \neq t$. If $F_s \neq F_t$ then either $t \notin F_s$ or $s \notin F_t$, because $t \in F_s$ implies $F_s \supseteq F_t$. Suppose $t \notin F_s$. Then 1_{F_s} is an exponential function separating s and t.

Let now $F_s = F_t$. Then $s,t \in \Phi$ and let

$$a = st^{-1} \in Q(\Phi)$$

be an element of the group of quotients of Φ.

Then, since $a \neq e$ and a has infinite order, the group G(a) generated by a is isomorphic to Z with addition

$$G(a) = \{a^z, z \in Z\} .$$

Choose $\theta \in]0,1[$ and define an exponential function on G(a) as

$$\chi_\theta(a^z) = \theta^z, \ z \in Z .$$

But any exponential function on a subgroup on a group can be extended to an exponential function on the whole group (Clifford and Preston (1961), Theorem 5.57) such that we can let χ_θ^* denote this extension.

Noting now that for any $u \in S$

$$us \in \Phi \Leftrightarrow F_{us} = F_s \Leftrightarrow u \in F_s$$

we can define

$$\chi(u) = \begin{cases} \dfrac{\chi_\theta^*(\psi(us))}{\chi_\theta^*(\psi(s))} & \text{if } us \in \Phi \\ 0 & \text{otherwise,} \end{cases}$$

where $\psi : \Phi \to Q(\Phi)$ is the canonical imbedding. χ is then a character on S separating s and t. □

A semigroup is <u>graduated</u> by τ if τ is a homomorphism of S into the non-negative integers $\mathbb{N}$ with addition.

If S is a semigroup and G is a set generating S we can construct a graduated semigroup by letting

$$\tilde{S} = \dot{\bigcup_{n \in \mathbb{N}}} G^n \text{ (disjoint union)} ,$$

defining the obvious composition and using $\tau(\tilde{s}) = n$ for $\tilde{s}$ in the canonical image of G^n.

This shall be called the <u>G-graduation</u> of S.

A (positive) measure μ on S is given by a function $\mu: S \to [0,\infty]$ such that for $A \subset S$

$$\mu(A) = \sum_{s \in A} \mu(s).$$

(Here we have allowed us not to distinguish between μ as a set function and a point function).

μ is <u>locally finite</u> if $\mu(s) < \infty$ for all $s \in S$. μ is <u>finite</u> if $\mu(S) < \infty$. If μ is finite, the set $\{s : \mu(s) \neq 0\}$ is at most countable. If $\mu(S) = 1$, μ is a <u>probability</u>.

The <u>convolution</u> of two measures μ and ν is given as

$$\mu * \nu(s) = \sum_{\substack{(s_1,s_2) \in S \times S \\ s_1 s_2 = s}} \mu(s_1)\mu(s_2) .$$

The measures $M(S)$ form a semigroup under convolution of which the finite measures is a subsemigroup.

We shall call a semigroup S <u>proper</u> if the set $\{(s_1,s_2): s_1 s_2 = s\}$ is finite for all s. If S is proper, the locally finite measures form a subsemigroup under convolution.

For a measure μ we define its <u>Laplace transform</u> $\hat{\mu}$ as the function

$$\hat{\mu} : \mathrm{EXP}(S) \to [0,\infty]$$

given as

$$\hat{\mu}(\xi) = \sum_{t \in S} \xi(t)\mu(t) .$$

The Laplace transform is a "semialgebra-homomorphism" in the sense that

$$\widehat{(a\mu + b\nu)} = a\hat{\mu} + b\hat{\nu}$$

$$\widehat{\mu * \nu} = \hat{\mu} \cdot \hat{\nu}$$

for all $\mu,\nu \in \mathcal{M}(S)$, $a,b \in [0,\infty]$.

It follows from the results of Hewitt and Zuckermann (1956) that the Laplace transform, restricted to the finite measures, is injective if the <u>bounded</u> exponential functions separate points (loc.cit., Theorems 3.3 and 3.4).

1. Extremal families for semigroup statistics

We shall here consider a system $(X_n, n \in \mathbb{N})$ of spaces where $X_n = X$ for all n and where X is discrete and at most countable.

Further we consider a system of statistics

$$t_n : X^n \to \mathcal{Y}_n$$

given as

$$t_n(x_1,\ldots,x_n) = t(x_1)t(x_2)\ldots t(x_n),$$

where $t : X \to S$ is a function from X into an Abelian semigroup.

Of course, $\mathcal{Y}_n$ is defined as $t(X^n)$ and we shall assume that $t(X)$ generates S, i.e. that $S = \bigcup_n \mathcal{Y}_n$.

This sequence is trivially algebraically transitive, so to be able to speak about the extremal family that it generates, we have to assume that the structure functions

$$f_n(s) = |\{\underset{\sim}{x}_n \in X^n : t_n(\underset{\sim}{x}_n) = s\}|$$

are finite. This will be the case if S is proper and if $f_1(s) = f(s) < \infty$.

Then we obviously have

$$f_n(s) = f^{*n}(s), \quad \mathcal{Y}_n = \{s : f_n(s) > 0\}$$

the n-fold convolution of f, considered as a locally finite measure.

We can then as in section 7, ch. II speak about the extremal family generated by the sequence of statistics t_n, $n \in \mathbb{N}$. We shall here try to identify this family somewhat more explicitly and see that it deserves the name of an exponential family.

We construct the Q-system as the system of <u>microcanonical distributions</u>.

$$Q_n^y(\underset{\sim}{x}_n) = \frac{1}{f_n(y)} \chi_{t_n^{-1}(y)}(\underset{\sim}{x}_n) .$$

Elementary calculations show that the ${}_mQ_n$-matrices then will be given as

$${}_mQ_n(s_m, s_n) = \frac{f^{*m}(s_m) \sum_{x:xs_m=s_n} f^{*(n-m)}(x)}{f^{*n}(s_n)} .$$

As we know from chapter II, the extremal family is given by the extreme points of the convex set of sequences $(\mu_n)_{n\in\mathbb{N}}$ of probability measures on S satisfying

$${}_mQ_n\mu_n = \mu_m.$$

This is easily seen to be equivalent to finding the extreme points of the convex set of sequences of non-negative functions $(h_n)_{n\in\mathbb{N}}$, satisfying

$$\sum_t \sum_{x:xs=t} f^{*k}(x) h_{n+k}(t) = h_n(s)$$

or, equivalently

$$\sum_x f^{*k}(x) h_{n+k}(xs) = h_n(s), \tag{1.1}$$

and further

$$\sum_s f^{*n}(s) h_n(s) = 1, \tag{1.2}$$

Where we have let $h_n(s) = \frac{\mu_n(s)}{f^{*n}(s)}$.

To find the extreme points of this set H, we shall need a lemma:

<u>1.1 Lemma</u> <u>For</u> $h \in H$ and $h_k(x) > 0$, <u>the sequences</u> $T_{x,k}h$, $x \in Y_k$, $k \in \mathbb{N}$ <u>are all in</u> H <u>where</u>

$$(T_{x,k}h)_n(s) = h_{n+k}(sx)/h_k(x).$$

<u>Proof</u> We just have to verify (1.1) and (1.2) for $T_{x,k}h$, knowing that it holds for h.

$$\sum_y f^{*\ell}(y)(T_{x,k}h)_{n+\ell}(ys) =$$

$$\sum_y f^{*\ell}(y)h_{n+k+\ell}(yxs)/h_k(x) = h_{n+k}(xs)/h_k(x),$$

which proves (1.1). Further

$$\sum_s f^{*n}(s)h_{n+k}(sx)/h_k(x) = h_k(x)/h_k(x) = 1$$

by (1.1), which proves (1.2). □

We are now ready to prove the following

<u>1.2 Proposition</u> <u>The extreme points of the convex set</u> H <u>are sequences satisfying</u>

$$h_n(s)h_k(x) = h_{n+k}(sx) \tag{1.3}$$

<u>for all</u> $s \in Y_n$, $x \in Y_k$.

<u>Proof</u> Let h be an extreme point of H. Rewrite (1.1) as

$$h_n(s) = \sum_x h_k(x)f^{*k}(x)(T_{x,k}h)_n(s).$$

By (1.2) this equation expresses h as a convex combination of $T_{x,k}h$, $x \in Y_k$. Because h is an extreme point, we must have

$$h_n(s) = (T_{x,k}h)_n(s) \quad \text{if } h_k(x)f^{*k}(x) > 0$$

or, equivalently

$$h_n(s)h_k(x) = h_{n+k}(sx)$$

if $s \in \mathcal{Y}_n$, $x \in \mathcal{Y}_k$ and $h_k(x) > 0$.

But, $h_k(x) = 0 \Rightarrow h_{n+k}(sx) = 0$ by (1.1) such that this equation holds for all $s \in \mathcal{Y}_n, x \in \mathcal{Y}_k$.

Suppose now that h is of this form. We want to show that h is extreme. By the integral representation, we can always write h as a mixture of extreme h' - s, i.e.

$$h_n(s) = \int_{h^* \in E} h_n^*(s)\mu(dh^*) \quad .$$

But then, for all n and s

$$h_{2n}(s+s) = (h_n(s))^2 = (\int h_n^*(s)\mu(dh^*))^2$$

$$= \int h_{2n}^*(s+s)\mu(dh^*)$$

$$= \int (h_n^*(s))^2\mu(dh^*) \quad .$$

But this implies $h_n^*(s) = h_n(s)$ a.s. μ, which proves that h is extreme. □

An alternative proof of the extremality of h could be based on the Hewitt-Savage 0-1 law (Hewitt and Savage (1955)):

According to P_h, $X_1,\dots,X_n$ are i.i.d. with density

$$P_h(X_1 = x) = h_1(t(x)) \quad .$$

Thus, the exchangeable σ-algebra of $(X_1,\dots,X_n,\dots)$ is P_h-trivial.

Since the tail σ-algebra of $t_n(\underset{\sim}{X}_n)$, $n \in \mathbb{N}$ is contained in the exchangeable, it must also be P_h-trivial and h is thus extreme.

On the other hand, as we shall see later, the Hewitt-Savage 0-1 law can be deduced from the above result, and it is therefore of interest to notice that the result is proved independently of the 0-1 law.

If the semigroup S is sufficiently well behaved, equation (1.3) will imply that

$$h_n(s) = \theta^n \xi(s) \tag{1.4}$$

for some $\xi \in \mathrm{EXP}(S)$ and this will again imply that

$$\theta^{-1} = \Sigma \; f(s)\xi(s) = \hat{f}(\xi) \; .$$

But in fact, this is no genuine restriction, since we can consider the <u>graduated semigroup</u> by

$$\tilde{S} = \dot{\bigcup_{n \in \mathbb{N}}} \mathcal{Y}_n \; ,$$

and we might then without loss of generality consider t as taking values in $\tilde{S}$. Equation (1.3) then says that (1.4) holds even with $\theta = 1$.

Note that, if in fact $t(\underset{\sim}{X}_k) = S$ for some k, we can define

$$\xi(s) = \frac{h_k(sx_0)}{h_k(x_0)} \; , \quad \theta = \frac{h_{k+1}(x_0)}{h_k(x_0)}$$

and get immediately that

$$\theta^n \xi(s) = \left(\frac{h_{k+1}(x_0)}{h_k(x_0)}\right)^n \frac{h_k(sx_0)}{h_k(x_0)} = \frac{h_{(n+1)k+n}(sx_0^n)}{h_{(n+1)k}(x_0^n)} = h_n(s) \; ,$$

such that we in this case also have the form (1.3) for all functions satisfying (1.4).

<u>In the following, we will therefore assume, without explicitly mentioning this fact, that S is such that</u> (1.3) <u>implies</u> (1.4).

As a corollary, we get:

<u>1.3 Proposition</u> <u>The extremal family generated by the sequence of semigroup statistics</u> $(t_n)_{n \in \mathbb{N}}$, <u>is the exponential family</u> $\{P_\xi, \xi \in \mathcal{D}\}$, <u>where</u>

$$P_\xi(x_1,\ldots,x_n) = \frac{1}{\hat{f}(\xi)^n}\,\xi(t(x_1)\ldots t(x_n)) = \prod_{i=1}^{n}\left(\frac{\xi(t(x_i))}{\hat{f}(\xi)}\right),$$

<u>and</u>

$$\mathcal{D} = \{\xi \in \mathrm{EXP}(S) : 0 < \hat{f}(\xi) = \sum_s f(s)\xi(s) < \infty\} .$$

<u>Proof</u> This follows directly from Proposition 1.2 and the fact that the correspondence $h \leftrightarrow \mu$ is one-to-one, onto and linear. □

Using our limit theorems from chapter II we get that with P_ξ-probability one

$${}_mQ_N(s_m,t_N(X_N)) = \frac{f^{*m}(s_m)\displaystyle\sum_{s:ss_m=t_N(X_N)} f^{*(N-m)}(s)}{f^{*N}(t_N(X_N))} \underset{N\to\infty}{\to} f^{*m}(s_m)\,\frac{\xi(s_m)}{\hat{f}(\xi)^m},$$

which can be thought of as an approximation formula for the conditional distributions. The approximation is not extremely useful in this form, but we shall later see how it can be made much more precise if we assume more about the semigroup S.

Clearly, the exponential families introduced here are different from the "usual" exponential families.

There are no dimensionality restrictions, the exponential functions are defined somewhat differently which, among other things, imply that the <u>support</u> of the P_ξ-measures depends heavily on ξ. As mentioned in sec. 0, the filters of S are characterised as the regions where exponential functions are positive. Thus in general, all the different filters F of S will correspond to subfamilies of $\{P_\xi, \xi \in \mathcal{D}\}$ that have support exactly on $t^{-1}(F \cap \mathcal{Y}_1)$.

As we shall see when we discuss inference in these families, this is a convenience rather than a nuisance.

2. General exponential families

Motivated by the results of the previous section we introduce the notion of a general exponential family.

Let α be a positive measure on the discrete, at most denumerable set X, such that $\alpha(x) > 0$ for all $x \in X$.

Let S be a semigroup and $t : X \to S$ a statistic. Assume that $t(X)$ generates S and that the transformed measure on S

$$\nu(s) = \sum_{x \in t^{-1}(s)} \alpha(x)$$

has finite convolution powers of any order.

A family P_1 of probability measures on X is said to be <u>general exponential</u> if

$$P_1 = \{P_\xi, \xi \in \mathcal{D}\},$$

where

$$P_\xi(x) = \alpha(x) \frac{\xi(t(x))}{\hat{\nu}(\xi)}$$

and

$$\mathcal{D} = \{\xi \in \mathrm{EXP}(S) : 0 < \hat{\nu}(\xi) < \infty\} .$$

As we see, such a family differs from those considered in sec. 1 by introducing the measure α and changing the assumption of S being proper to that of ν having finite convolution powers. For $\alpha(x) \equiv 1$, i.e. counting measure, we get the families considered before.

Consider now the following repetitive structure

$$\underset{\sim}{X}_n = X^n, \quad t_n(x_1,\ldots,x_n) = t(x_1)\ldots t(x_n)$$

and consider the family P on $\underset{\sim}{X}$ given by the specification that $X_1,\ldots$ should be independent, identically distributed with distributions in P_1.

<u>2.1 Proposition</u> <u>A general exponential family is extremal in the above specified repetitive structure.</u>

<u>Proof</u> The technical part of this proof is identical to that of Proposition 1.3 and is therefore omitted. We shall here just indicate that one defines

$$Y_n = t_n(\underset{\sim}{X}_n)$$

and verifies that

$$_mQ_n(s_m,s_n) = \frac{\nu^{*m}(s_m) \sum_{s:ss_m=s_n} \nu^{*(n-m)}(s)}{\nu^{*n}(s_n)}, \tag{2.1}$$

whereby the proof only differs from the cited by the minor change that f everywhere should be substituted with ν. □

Again we can use the general limit theorem and get an "approximation" for the conditional distributions as

$$_mQ_N(s_m,s_N) \sim \nu^{*m}(s_m) \frac{\xi(s_m)}{\hat{\nu}(\xi)^m} \tag{2.2}$$

for an appropriate choice of ξ and large N.

The families considered above look a bit more like the "usual" exponential families. Comparing the expressions

$$P_\xi(x) = \alpha(x) \frac{\xi(t(x))}{\hat{\nu}(\xi)}$$

and

$$P(x|\theta) = a(x) \frac{e^{\theta \cdot t(x)}}{\phi(\theta)} ,$$

we see that α plays the role of a, $t \to \xi(t)$ that of the function $e^{\theta \cdot t}$ and the Laplace transform $\hat{\nu}(\xi)$ that of ϕ. Apart from that there are still differences present of the type discussed in the previous section.

3. The classical case. Z^d-valued statistics

We shall here assume that the semigroup in question is a subsemigroup of Z^d. We might then as well assume that d is minimal with this property.

The strictly positive exponential functions on Z^d are given as

$$z = (z_1,\dots,z_d) \to \theta_1^{z_1}\dots\theta_d^{z_d} = \theta^z$$

for $\theta = (\theta_1,\dots,\theta_d) \in \mathbb{R}_+^d$.

We can assume that we for each non-empty filter F have an injective homomorphism

$$\phi^F : F \to Z^{d(F)}$$

$$\phi^F(s) = (\phi_1^F(s),\dots,\phi_{d(F)}^F(s))$$

and correspondingly that the exponential functions of S that are positive exactly on F are given as

$$s \to \theta_1^{\phi_1^F(s)}\dots\theta_{d(F)}^{\phi_{d(F)}^F(s)} = \theta_F^{\phi^F(s)}$$

where $\theta_F = (\theta_1,\dots,\theta_{d(F)}) \in \mathbb{R}_+^{d(F)}$.

We thus get a complete parametrisation of the exponential functions by

$$s \to \theta^{\phi^F(s)}, \quad F \in \mathcal{F}(S), \quad \theta_F \in \mathbb{R}_+^{d(F)},$$

or, by taking logarithms,

$$s \to e^{\langle\theta_F,\phi^F(s)\rangle}, \quad F \in \mathcal{F}(S), \quad \theta_F \in \mathbb{R}^{d(F)},$$

where $\langle\cdot,\cdot\rangle$ denotes Euclidean inner product on $\mathbb{R}^{d(F)}$.

Returning now to the exponential families we see that a generalised exponential family in the classical case has the form

$$\mathcal{P} = \{P_{\xi,F}, \xi \in \mathcal{D}_F, F \in \mathcal{F}(S), F \neq \emptyset\}$$

where

$$P_{\xi,F}(x) = \alpha(x) \frac{e^{\langle\xi,\phi^F(s(x))\rangle}}{\psi_F(\xi)}$$

$$\psi_F(\xi) = \sum_x \alpha(x) e^{\langle\xi,\phi^F(s(x))\rangle}$$

$$\mathcal{D}_F = \{\xi \in \mathbb{R}^{d(F)} : \psi_F(\xi) < \infty\}$$

and $\phi^F : S \to Z^{d(F)}$ are fixed but arbitrary isomorphisms of F onto a subsemigroup of $Z^{d(F)}$.

Hence we see that a generalised exponential family in this case is <u>a family of</u> usual exponential families, one for each non-empty filter $F \in \mathcal{F}(S)$. Each of these has as canonical statistic $\phi^F \circ s$.

For families of the above type we can give a much more precise meaning to the limit statement about the conditional distributions ${}_mQ_n$. This shall be derived via a local limit theorem that in its present form is due to Per Martin-Löf.

We consider a family of probabilities on Z^d of the form

$$p(t,\theta) = \frac{a(t)}{\phi(\theta)} e^{\langle\theta,t\rangle}, \quad \theta \in D$$

$$D = \{\theta \in \mathbb{R}^d : \phi(\theta) = \sum_{t\in Z^d} a(t) e^{\langle\theta,t\rangle} < \infty\},$$

i.e. a usual 'reduced' exponential family. We assume that d is chosen minimal in the sense that the distributions are not concentrated on any set of the form $t+H$, where H is a proper subgroup of Z^d. This in turn implies that the characteristic function for $p(\cdot,\theta)$ is not equal to 1 except at zero. Further that the covariance matrix of t is not singular but has full rank $=d$.

We define

$$\xi_\theta = D\log\phi(\theta) = E_\theta(T)$$

$$V_\theta = D^2\log\phi(\theta) = V_\theta(T)$$

$$\mu_\theta = D^3\log\phi(\theta) \ .$$

ξ_θ, V_θ and μ_θ are the first three cumulants in the distribution $p(\cdot,\theta)$.

If $T_1,\ldots,T_n$ are i.i.d.r.v. with common distribution $p(\cdot,\theta)$, their sum $T^{(n)} = T_1+\ldots+T_n$ has distribution

$$p^{(n)}(t,\theta) = p^{*n}(t,\theta) = \frac{a^{*n}(t)}{\phi(\theta)^n}\, e^{\langle\theta,t\rangle}$$

and for the terms above we have

$$\xi_\theta^{(n)} = n\xi_\theta,\ V_\theta^{(n)} = nV_\theta,\ \mu_\theta^{(n)} = n\mu_\theta \ .$$

Define

$$u_n = V_\theta^{-\frac{1}{2}}\left(\frac{t-n\xi_\theta}{\sqrt{n}}\right) \ .$$

We then have the following

3.1 Theorem (Local limit theorem).

$$n^{d/2}\sqrt{\det V_\theta}\; p^{(n)}(t,\theta)$$

$$= \frac{1}{(2\pi)^{d/2}} e^{-\frac{\|u_n\|^2}{2}} \left(1 + \frac{1}{\sqrt{n}} P_\theta(u_n)\right) + 0\left(\frac{1}{n}\right),$$

where P_θ is a polynomial of degree 3 in u_n with coefficients that depend continuously on θ. The approximation is uniform in (t,θ) for $t \in Z^d$ and $\theta \in K$ for any compact subset K of the interior of D. $P_\theta(0) = 0$.

Proof We shall just consider the case with $d = 1$ to ease the notational problem. The proof is not essentially different for $d > 1$.

The relation to be proved for $d = 1$ reduces to

$$\sqrt{nV_\theta}\; p^{(n)}(t,\theta) = \frac{1}{\sqrt{2\pi}} e^{\frac{-u_n^2}{2}} \left(1 + \frac{\mu_\theta}{6V_\theta^{3/2}\sqrt{n}}(u_n^3 - 3u_n)\right) + 0\left(\frac{1}{n}\right), \qquad (*)$$

where we also have given the explicit form of the polynomial.

Let $\psi_\theta(\lambda)$ denote the characteristic function for T_1. The characteristic function for U_1 is then

$$\gamma_\theta(\lambda) = Ee^{i\lambda\left(\frac{T_1 - \xi_\theta}{\sqrt{V_\theta}}\right)} = \psi_\theta(\lambda)e^{-i\xi_\theta\frac{\lambda}{\sqrt{V_\theta}}}.$$

For λ sufficiently small (independently of θ when $\theta \in K$), we have

$$\log\gamma_\theta(\lambda) = -\frac{1}{2}\lambda^2 + \frac{i^3}{3!}\frac{\mu_\theta}{V_\theta^{3/2}}\lambda^3 + R_\theta(\lambda)$$

where

$$|R_\theta(\lambda)| \leqq C|\lambda|^4 \quad \text{for } \theta \in K. \tag{3.1}$$

This follows from the fact that $\log\gamma_\theta$ is the cumulant generating function.

The characteristic function for U_n is

$$\gamma_{\theta,n}(\lambda) = (\gamma_\theta(\frac{\lambda}{\sqrt{n}}))^n ,$$

so that

$$\log\gamma_{\theta,n}(\lambda) = -\frac{1}{2}\lambda^2 + \frac{i^3}{3!}\frac{\mu_\theta}{\sqrt{n}V_\theta^{3/2}}\lambda^3 + n\,R_\theta(\frac{\lambda}{\sqrt{n}}).$$

By Fourier inversion we get

$$\sqrt{nV_\theta}\; p^{(n)}(t,\theta) = \frac{1}{2\pi}\int_{|\lambda|<\pi\sqrt{nV_\theta}} e^{-iu_n\lambda}\gamma_{\theta,n}(\lambda)\,d\lambda.$$

If we differentiate three times w.r.t. u in the expression

$$\frac{1}{\sqrt{2\pi}}e^{\frac{-u^2}{2}} = \frac{1}{2\pi}\int_{-\infty}^{\infty} e^{-iu\lambda}e^{\frac{-\lambda^2}{2}}d\lambda ,$$

we get

$$\frac{1}{\sqrt{2\pi}}e^{\frac{-u^2}{2}}(u^3-3u) = \frac{1}{2\pi}\int_{-\infty}^{\infty} e^{-iu\lambda}e^{\frac{-\lambda^2}{2}}(i\lambda)^3\,d\lambda$$

such that the main term of the right-hand side in (*) becomes equal to

$$\frac{1}{2\pi}\int_{-\infty}^{\infty} e^{-iu\lambda}\, e^{\frac{-\lambda^2}{2}}\left(1+\frac{i^3}{3!}\,\frac{\mu_\theta}{\sqrt{n}V_\theta^{3/2}}\lambda^3\right)d\lambda$$

$$= \frac{1}{2\pi}\int_{-\infty}^{\infty} e^{-iu\lambda}\delta_{\theta,n}(\lambda)\,d\lambda.$$

The difference between the LHS and the main term of the RHS is thus

$$\frac{1}{2\pi}\int_{-\infty}^{\infty} e^{-in\lambda}(\tilde{\gamma}_{n,\theta}(\lambda) - \delta_{\theta,n}(\lambda))\,d\lambda,$$

where $\tilde{\gamma}_{n,\theta}(\lambda) = \gamma_{n,\theta}(\lambda)$ for $|\lambda| < \pi\sqrt{nV_\theta}$ and 0 otherwise.

This difference is less in absolute value than

$$\frac{1}{2\pi}\int_{|\lambda|<\delta\sqrt{n}} |\gamma_{n,\theta}(\lambda) - \delta_{\theta,n}(\lambda)|\,d\lambda + \frac{1}{2\pi}\int_{|\lambda|>\delta\sqrt{n}} (|\tilde{\gamma}_{n,\theta}(\lambda)|)+|\delta_{\theta,n}(\lambda)|\,d\lambda$$

$$= I_1 + I_2,$$

where δ is a constant to be chosen later.

We first give a <u>bound for I_1</u>:

For δ sufficiently small ($\delta \leqq \delta_0$) we have

$$I_1 = \frac{1}{2\pi}\int_{|\lambda|<\delta\sqrt{n}} e^{-\frac{\lambda^2}{2}}\left|\left(e^{\frac{i^3}{3!}\frac{\mu_\theta}{\sqrt{n}V_\theta^{3/2}}\lambda^3 + nR_\theta(\frac{\lambda}{\sqrt{n}})} - 1 - \frac{i^3}{3!}\,\frac{\mu_\theta}{\sqrt{n}V_\theta^{3/2}}\lambda^3\right)\right|d\lambda.$$

If we use the inequality

$$|e^{x+y} - 1 - x| \leqq |y| + \frac{|x+y|^2}{2}e^{|x+y|},$$

that easily can be shown via a series expansion of the exponential function, we get that the function under the integral sign is less than

$$e^{-\frac{\lambda^2}{2}}\left(|nR_\theta(\tfrac{\lambda}{\sqrt{n}})| + \frac{|nR_\theta^*(\frac{\lambda}{\sqrt{n}})|^2}{2} e^{|nR_\theta^*(\frac{\lambda}{\sqrt{n}})|}\right) ,$$

where

$$R_\theta^*(\lambda) = \frac{i^3}{3!}\frac{\mu_\theta}{V_\theta^{3/2}}\lambda^3 + R_\theta(\lambda) .$$

and

$$|R_\theta^*(\lambda)| \leqq C^*|\lambda|^3, \tag{3.2}$$

where C^* does not depend on θ for $\theta \in K$.

For $|\lambda| < \delta\sqrt{n}$ we clearly have

$$|nR_\theta^*(\tfrac{\lambda}{\sqrt{n}})| \leqq C^*\frac{|\lambda|^3}{\sqrt{n}\cdot n} \leqq C^*\delta\lambda^2.$$

If we now choose $\delta = \min(\delta_0, \frac{1}{4C^*})$, we have

$$|nR_\theta^*(\tfrac{\lambda}{\sqrt{n}})| \leqq \tfrac{1}{4}\lambda^2. \tag{3.3}$$

Using (3.1), (3.2) and (3.3), we get for the function under the integral sign in I_1 that this is less than

$$e^{-\frac{\lambda^2}{2}}\left(\tfrac{1}{n}C\lambda^4 + \tfrac{1}{2n}(C^*)^2\lambda^6 e^{\frac{1}{4}\lambda^2}\right)$$

$$= \tfrac{1}{n}\left(C\lambda^4 e^{-\frac{\lambda^2}{2}} + \frac{(C^*)^2}{2}\lambda^6 e^{-\frac{1}{4}\lambda^2}\right)$$

such that

$$I_1 \leq \frac{1}{2\pi n} \int_{|\lambda|<\delta\sqrt{n}} (C\lambda^4 e^{-\frac{\lambda^2}{2}} + \frac{C^*}{2}\lambda^6 e^{-\frac{\lambda^2}{4}})\, d\lambda$$

$$\leq \frac{C^{**}}{n} = O(\frac{1}{n}) \ ,$$

since the above function is integrable over $\mathbb{R}$.

To get a bound for I_2, we proceed as follows:

Substitute $\tau = \lambda/\sqrt{n}$ to get

$$I_2 = \frac{1}{2\pi} \int_{\delta<|\tau|<\pi} \sqrt{n}\, |\psi_\theta(\frac{\pi}{\sqrt{V_\theta}})|^n\, d\tau +$$

$$+ \frac{1}{2\pi} \int_{|\tau|>\delta} \sqrt{n}\, e^{-\frac{n\tau^2}{2}} \, |(1 + \frac{i^3}{3!} \frac{\mu_\theta}{V_\theta^{3/2}} n\tau^3)|\, d\tau$$

both these integrals decrease exponentially to zero since $|\psi_\theta(\frac{\tau}{\sqrt{V_\theta}})| \leq \rho < 1$ for $|\tau| > \delta$ because of the assumption of the theorem. We have now shown that the difference between the main term on the RHS and the LHS of (*) is $O(\frac{1}{n})$ and the proof is complete.

In the case of $d > 1$ the proof is identical apart from the fact that the Fourier inversion formula has to be modified accordingly and instead of V_θ, the smallest eigenvalue of V_θ occurs. □

3.2 Corollary For a function $a : Z^d \to \mathbb{R}^+$ with finite convolution powers, we have

$$a^{*n}(t) = \frac{\phi(\theta)^n e^{-\langle\theta,t\rangle}}{(2\pi n)^{d/2}\sqrt{\det V_\theta}} \cdot \Big[e^{-\frac{1}{2n}(t-n\xi_\theta)' V_\theta^{-1}(t-n\xi_\theta)}$$

$$\times \ (1 + \frac{1}{\sqrt{n}} P_\theta((nV_\theta)^{-1/2}(t-n\xi_\theta))) \ + O(\frac{1}{n}) \Big]$$

uniformly in t and for $\theta \in K$, where K is any compact subset of the interior of D and all terms are defined as in the Theorem.

Proof Just use that $p^{(n)}(t,\theta) = \frac{a^{*n}(t)}{\phi(\theta)^n} e^{<\theta,t>}$ and isolate $a^{*n}(t)$. □

Note that the left hand side of this approximation formula does not involve θ. We thus have a continuum of approximation formulae for $a^{*n}(t)$ and are able to choose θ conveniently, e.g. such that it varies with n in certain ways. If we for example choose $\theta = \hat{\theta}$ to satisfy the equation

$$t - n\xi_{\hat{\theta}} = 0,$$

we get

3.3 Corollary For a function $a : Z^d \to \mathbb{R}^+$ with finite convolution powers, we have

$$a^{*n}(t) = \frac{\phi(\hat{\theta})^n e^{-<\hat{\theta},t>}}{(2\pi n)^{d/2}\sqrt{\det V_{\hat{\theta}}}} \left(1 + O\left(\frac{1}{n}\right)\right),$$

uniformly for $\hat{\theta} \in K$ where K is a compact subset of the interior of D. (Note that $\hat{\theta}$ depends on t).

Proof Since $P_\theta(0) = 0$, we obtain the result by inserting $\hat{\theta}$ into the approximation formula from the first corollary. □

3.4 Example For $d = 1$ and $a(t) = 1_{\{0,1\}}(t)$ we have

$$a^{*n}(t) = \binom{n}{t} \text{ and } \phi(\theta) = (1 + e^{\theta}).$$

Further,

$$\xi_\theta = \frac{e^\theta}{1+e^\theta}\ ,\ \hat{\theta} = \log\ (\frac{t}{n-t})\ ,\ V_\theta = \frac{e^\theta}{(1+e^\theta)^2}$$

and thus

$$\binom{n}{t} = \frac{(1+\frac{t}{n-t})^{n+1}\ (\frac{n-t}{t})^t}{\sqrt{2\pi n}\ \sqrt{\frac{t}{n-t}}}(1+O(\frac{1}{n}))$$

$$= \frac{n^{n+\frac{1}{2}}}{\sqrt{2\pi}\ (n-t)^{n-t+\frac{1}{2}} t^{t+\frac{1}{2}}}(1+O(\frac{1}{n})).$$

This expression is exactly what we get by applying Stirlings formula to $\binom{n}{t}$.

<u>3.5 Example</u> For $a(t) = \frac{1}{t!}$, we have $a^{*n}(t) = \frac{n^t}{t!}$, $\phi(\theta) = e^{e^\theta}$, $\xi_\theta = V_\theta = e^\theta$. $\hat{\theta} = \log(\frac{t}{n})$. If we let $t = n$, we get

$$\frac{n^n}{n!} = \frac{1}{\sqrt{2\pi n}}\ e^n(1+O(\frac{1}{n}))$$

which is equivalent to

$$n! = \sqrt{2\pi}\ n^{n+\frac{1}{2}}\ e^{-n}(1+O(\frac{1}{n}))$$

and this <u>is</u> Stirlings formula.

Using this theorem and the corollaries we can give (2.2) a much more precise meaning.

In the classical case we namely have that (2.1) simplifies to

$${}_nQ_N(t,t_N) = \frac{\gamma^{*n}(t)\gamma^{*(N-n)}(t-t_N)}{\gamma^{*N}(t)}$$

where $t = \phi_F \circ s_n(\underset{\sim}{x}_n)$.

We can now use the corollaries on the function (measure) $\gamma^{*n}(t)$ such that we get

<u>3.6 Proposition</u> <u>The conditional distributions</u> ${}_nQ_N$ <u>satisfy for</u> $N \to \infty$

$$ {}_nQ_N(t,t_N) = \gamma^{*n}(t_n)\, \frac{e^{<\hat{\theta}_N,t>}}{\phi(\hat{\theta}_N)^n}\, (1+O(\tfrac{1}{N})) \;, $$

<u>for any fixed but arbitrary</u> $F \in F(S)$, $F \neq \emptyset$, $t \in \phi_F \circ s_n(\underset{\sim}{X}_n)$. The <u>approximation is uniform for</u> $t_N \in F$ <u>with</u> $\theta_N = \hat{\theta}(t_N) \in K$, <u>where</u> K <u>is any compact subset of the interior of</u> $\mathcal{D}_F$.

<u>Proof</u> We use the approximation in Corollary 3.2 for $\gamma^{*(N-n)}$ and that in Corollary 3.3 for γ^{*N}. Thus we let $\hat{\theta}_N = \hat{\theta}(t_N)$ satisfy $t_N = N\xi_{\hat{\theta}_N}$. This then implies

$$ (t_N - t) - (N-n)\xi_{\hat{\theta}_N} = n\xi_{\hat{\theta}_N} - t. $$

Since $P_\theta(0) = 0$, P_θ does not have a constant term and

$$ P_{\hat{\theta}_N}\left(\frac{1}{\sqrt{N-n}}\, V_{\hat{\theta}_N}^{-\frac{1}{2}}(n\xi_{\hat{\theta}_N} - t)\right) = O\left(\frac{1}{\sqrt{N}}\right). $$

Further,

$$ e^{-\frac{1}{2(N-n)}\,(n\xi_{\hat{\theta}_N}-t)'V_{\hat{\theta}_N}^{-1}(n\xi_{\hat{\theta}_N}-t)} = (1+O(\tfrac{1}{N})) $$

because $(n\xi_{\hat{\theta}} - t)'V_{\hat{\theta}_N}^{-1}(n\xi_{\hat{\theta}_N} - 1)$ is bounded for $\hat{\theta}_N \in K$.

Thus we get

$$\gamma^{*(N-n)}(t_N - t) = \frac{\phi(\hat{\theta}_N)^{N-n} e^{-<\hat{\theta}_N, t_N - t>}}{(2\pi(N-n))^{d(F)/2} \sqrt{\det V_{\hat{\theta}_N}}} \left[1 + O(\tfrac{1}{N})\right],$$

and

$$\gamma^{*N}(t_N) = \frac{\phi(\hat{\theta}_N)^N e^{-<\hat{\theta}_N, t_N>}}{(2\pi N)^{d(F)/2} \sqrt{\det V_{\hat{\theta}_N}}} (1 + O(\tfrac{1}{N})).$$

Inserting these into (3.4) we get

$${}_nQ_N(t,t_N) = \frac{\gamma^{*n}(t)\ e^{<\hat{\theta}_N, t>}}{\phi(\hat{\theta}_N)} \ (1 + O(\tfrac{1}{N})),$$

which was to be proved. □

Note that the law of large numbers implies $\hat{\theta}_N \to \theta$ with $p(\cdot,\theta)$ - probability one such that the above result implies

$${}_nQ_N(t,T_N) \to p^{(n)}(t,\theta) \ \ p(\cdot,\theta) \ \text{a.s.} \ ,$$

i.e. the result from Theorem 4.4 of § II. The present result is stronger in the sense that we also get the rate of convergence and a rather explicit form for $p^{(n)}(t,\theta)$. On the other hand Theorem 4.4 applies for <u>any</u> extremal family.

P. Diaconis and D. Freedman have in a series of papers discussed approximation of quantities like ${}_nQ_N(t,T_N)$ with $p^{(n)}(t,\theta)$, giving sharp bounds on the total variation distance between these. The bounds involve n,N and a constant. Such results are typically stronger but also require more structure. See Diaconis (1977) and Diaconis and Freedman (1980a, 1987a, 1987b) for further details.

4. Maximum likelihood estimation in general exponential families

We shall here give a short review of the most basic results about statistical inference in generalised exponential families. We want to convey the idea, that most known results about inference in 'usual' exponential families carry through to generalised exponential families, indeed sometimes these results become more streamlined than usual. We also want to indicate the points where this is not the case.

So, consider a generalised exponential family of distributions on $\mathcal{X}$

$$p(x,\theta) = \alpha(x)\,\frac{\theta(t(x))}{\phi(\theta)}\ , \quad \theta \in \mathcal{D}$$

where $t : \mathcal{X} \to S$ is a map from $\mathcal{X}$ into a discrete semigroup,

$$\phi(\theta) = \sum_x \alpha(x)\theta(t(x))\ , \quad \text{and}$$

$$\mathcal{D} = \{\theta \in \mathrm{EXP}(S) : 0 < \phi(\theta) < \infty\}$$

We shall further suppose that $\mathcal{D}$ separates points in S.

This assumption can be made without loss of generality since we otherwise define an equivalence relation $\sim$ on S by

$$s \sim s' \text{ iff } \frac{\theta(s)}{\phi(\theta)^n} = \frac{\theta(s')}{\phi(\theta)^{n'}},\ \forall\ \theta \in \mathcal{D}\ .$$

The equivalence relation is easily seen to be a congruence on S such that we can define

$$S^* = S/\sim\ ,\ t^* = p^* \circ t\ ,$$

where

$$p^* : S \to S^*$$

is the canonical map. We can then rewrite the probabilities as

$$p(x,\theta) = \alpha(x) \frac{\theta^*(t^*(x))}{\phi(\theta^*)} , \quad \theta^* \in \mathcal{D}^*$$

and this representation will have the required property. From section 0 we know that this implies that all filter components are cancellative etc.

So suppose now that we have observations $X_1 = x_1, \ldots, X_n = x_n$, where $X_i, i = 1,\ldots,n$ are i.i.d. r.v. with distribution $p(\cdot,\theta)$, $\theta \in \mathcal{D}$, θ unknown.

The likelihoodfunction becomes

$$L(\theta, \underset{\sim}{x}_n) = \frac{1}{\phi(\theta)^n} \prod_{i=1}^{n} \theta(t(x_i)) \prod_{i=1}^{n} \alpha(x_i)$$

$$= \frac{1}{\phi(\theta)^n} \theta(t_n(\underset{\sim}{x}_n)) \prod_{i=1}^{n} \alpha(x_i) \tag{4.1}$$

where $t_n(\underset{\sim}{x}_n) = t(x_1)\ldots t(x_n)$.

We shall show that the maximum likelihood estimate $\hat{\theta}$ of θ always exists, or more precisely

4.1 Proposition For all $\underset{\sim}{x}_n \in X^n$ such that $p(\underset{\sim}{x}_n,\theta) > 0$ for some θ, there is a $\hat{\theta}$, maximising (4.1), i.e. such that

$$L(\hat{\theta}, \underset{\sim}{x}_n) = \sup_{\theta \in \mathcal{D}} L(\theta, \underset{\sim}{x}_n) .$$

Proof The proof of this result involves a compactness argument, so we first prove a lemma:

<u>4.2 Lemma</u> <u>The set of subprobabilities</u> $P^* = \{cp(\cdot,\theta)\,|\,\theta \in \mathcal{D}, 0 \leqq c \leqq 1\}$ <u>is compact in the pointwise topology</u>.

<u>Proof of Lemma</u>: The set of subprobabilities is compact so the above set is compact iff it is closed. Thus let $(c_n, \theta_n)_{n \in \mathbb{N}}$ be a sequence such that

$$c_n\, p(x, \theta_n) \xrightarrow[n \to \infty]{} p(x) \;,$$

where $p(x) \geqq 0$, $\sum_x p(x) \leqq 1$. If $p(x) \equiv 0, p = 0 \cdot p(\cdot, \theta)$ and there is nothing to prove. Otherwise let

$$a_k^n(t) = \left(\frac{c_n}{\phi(\theta_n)}\right)^k \theta_n(t) \quad \text{for} \quad t \in t(X)^k.$$

Then we have for $t = t(x_1)\ldots t(x_k)$ that $a_k^n(t)$ is convergent for $n \to \infty$ with the limit

$$a_k(t) = \lim_{n\to\infty} a_k^n(t) = \frac{p(x_1)\ldots p(x_k)}{\alpha(x_1)\ldots\alpha(x_k)} \;.$$

Clearly, we have for $s \in t(X)^k$, $t \in t(X)^\ell$ that

$$a_k^n(s)\, a_\ell^n(t) = a_{k+\ell}^n(st)$$

and therefore also in the limit

$$a_k(s)\, a_\ell(t) = a_{k+\ell}(st).$$

Recalling (p. 145) that we have assumed this to imply

$$a_k(t) = c^k \theta(t)$$

for some $c \geqq 0$ and $\theta \in EXP(S)$, we obtain

$$c_n\, p(x,\theta_n) \to c\ \alpha(x)\theta(t(x)) = p(x) \quad .$$

Since $p(x) > 0$ we have $c > 0$ and

$$\phi(\theta) = \frac{1}{c} \sum_x p(x) < \infty \ ,$$

such that if we let $c^* = c\phi(\theta)$, we have $p = c^* p(\cdot,\theta)$ and $c^* \leqq 1$ because $c^* = \sum_x p(x)$. The lemma is proved. □

<u>Proof of the Proposition</u>: Extend the likelihood function to be defined on the subprobabilities in the lemma

$$\tilde{L}(c,\theta,\underset{\sim}{x}_n) = c^n \prod \alpha(x_i) \frac{1}{\phi(\theta)^n} \theta(t_n(\underset{\sim}{x}_n)) \quad .$$

$\tilde{L}$ is clearly continuous on the compact set P^*, such that it attains its maximum for some $(\hat{c},\hat{\theta})$. But then $\hat{c} = 1$, because otherwise it would not be a maximum. The maximum over the larger set is clearly also maximum over the smaller. The Proposition has been proved. □

Next we prove that $\hat{\theta}$ is uniquely determined in the following sense:

<u>4.3 Proposition</u> <u>$\hat{\theta}$ is uniquely determined in the sense that if θ_1,θ_2 both maximise (4.1) then</u> $p(\cdot,\theta_1) = p(\cdot,\theta_2)$.

<u>Proof</u> Assume that

$$\frac{\theta_1(t(\underset{\sim}{x}_n))}{\phi(\theta_1)^n} = \frac{\theta_2(t(\underset{\sim}{x}_n))}{\phi(\theta_2)^n} = \sup_{\theta \in \mathcal{D}} \frac{\theta(t(\underset{\sim}{x}_n))}{\phi(\theta)^n} \quad .$$

For $\theta_1,\theta_2 \in \mathrm{EXP}(S)$, $\sqrt{\theta_1\theta_2}$ defined as

$$\sqrt{\theta_1\theta_2}\ (t) = \sqrt{\theta_1(t)\theta_2(t)} \ ,$$

is clearly also an exponential function.

We have further by Cauchy-Schwarz' inequality that

$$\phi(\sqrt{\theta_1\theta_2}) = \Sigma\ \alpha(x)\sqrt{\theta_1(t(x))\theta_2(t(x))}$$

$$\leqq \sqrt{\Sigma\alpha(x)\theta_1(t(x))} \cdot \sqrt{\Sigma\alpha(x)\theta_2(t(x))} = \sqrt{\phi(\theta_1)\phi(\theta_2)}\ ,$$

such that $\sqrt{\theta_1\theta_2} \in \mathcal{D}$. The inequality is strict unless $\theta_1(t(x)) = \lambda\theta_2(t(x))$ for some $\lambda > 0$.

Thus

$$\frac{\sqrt{\theta_1\theta_2}(t(\underset{\sim}{x}_n))}{\phi(\sqrt{\theta_1\theta_2})^n} \geqq \frac{\sqrt{\theta_1(t(\underset{\sim}{x}_n))}}{\sqrt{\phi(\theta_1)^n}} \frac{\sqrt{\theta_2(t(\underset{\sim}{x}_n))}}{\sqrt{\phi(\theta_2)^n}} = \sup_{\theta\in\mathcal{D}} \frac{\theta(t(\underset{\sim}{x}_n))}{\phi(\theta)^n} \quad .$$

The inequality cannot be strict by the definition of sup such that we must have

$$\theta_1(t(x)) = \lambda\theta_2(t(x)),\ \phi(\theta_1) = \lambda\phi(\theta_2)$$

and thus $p(\cdot,\theta_1) = p(\cdot,\theta_2)$. □

For 'usual' exponential families we know that the maximum likelihood estimate does not always exist, but if it exists, it is unique. Thus the results above seem slightly more streamlined. Another useful property of 'usual' exponential families is the fact that, sometimes, the maximum likelihood estimate can be found by equating the observed value of the canonical statistic to its expectation:

$$E_\theta t(X) = t(x_{obs}) \ .$$

We shall here derive a similar equation, but first we shall get rid of some small technical difficulties by noting

<u>4.4 Proposition</u> <u>If</u> $\hat{\theta}$ <u>maximises</u> (4.1) <u>then</u> $\{s \in S : \hat{\theta}(s) > 0\} = F_{t_n(\underset{\sim}{x}_n)}$, <u>the smallest filter of S containing the observation</u>.

Proof Let $F = F_{t_n(\underset{\sim}{x}_n)}$ be the smallest filter containing the observation. Since $F_{\hat{\theta}} = \{s : \hat{\theta}(s) > 0\}$ is a filter, it clearly contains F. But $\hat{\theta} \cdot 1_F \in \mathrm{EXP}(S)$ and

$$\phi(\hat{\theta}1_F) = \sum_{x \in t^{-1}(F)} \alpha(x)\hat{\theta}(t(x)) \leqq \phi(\hat{\theta})$$

and there is equality iff $\hat{\theta}1_F = \hat{\theta}$, i.e. $F_{\hat{\theta}} = F$.

But then

$$L(\hat{\theta}1_F, \underset{\sim}{x}_n) \geqq L(\hat{\theta}, \underset{\sim}{x}_n) \Rightarrow F_{\hat{\theta}} = F \ . \quad \square$$

This means that we only have to search for estimates among those θ's that are positive exactly on $F = F_{t(\underset{\sim}{x}_n)}$. For each filter F we now let

$$\mathcal{D}_F = \{\theta \in \mathcal{D} : F^\theta = F\} \ .$$

We have no immediate definition of an expectation on a semigroup, but for $\theta, \eta \in \mathcal{D}_F$ we can consider the quantity

$$m_\theta(\eta) = E_\theta \log \frac{\theta(t(X))}{\eta(t(X))} = \sum_{x \in t^{-1}(F)} \alpha(x) \frac{\theta(t(x))}{\phi(\theta)} \log \frac{\theta(t(x))}{\eta(t(x))} \ .$$

This quantity is well defined and satisfies an inequality because

$$m_\theta(\eta) = \frac{\phi(\eta)}{\phi(\theta)} \sum_x p(x;\eta) \frac{\theta(t(x))}{\eta(t(x))} \log \frac{\theta(t(x))}{\eta(t(x))} \ ,$$

and $u \to u \log u$ is convex and bounded below by $-e^{-1}$ such that the negative part of the sum converges and

$$m_\theta(\eta) \geqq \frac{\phi(\eta)}{\phi(\theta)} (\sum_x p(x,\eta) \frac{\theta(t(x))}{\eta(t(x))}) \cdot \log(\sum_x p(x,\eta) \frac{\theta(t(x))}{\eta(t(x))}) = \log\frac{\phi(\theta)}{\phi(\eta)} \quad . \tag{4.2}$$

We then have

<u>4.5 Proposition</u> If $\hat{\theta} \in \mathcal{D}_F$ <u>satisfies the equations</u>

$$\frac{1}{n}[\log \hat{\theta}(t(\underset{\sim}{x}_n)) - \log \eta(t(\underset{\sim}{x}_n))] = m_{\hat{\theta}}(\eta) \quad \forall \eta \in \mathcal{D}_F \ , \tag{4.3}$$

<u>then</u> $\hat{\theta}$ <u>is the maximum likelihood estimate of</u> θ, <u>based on</u> $\underset{\sim}{X}_n = \underset{\sim}{x}_n$.

<u>Proof</u> Using (4.2) we get

$$\frac{1}{n}[\log \hat{\theta}(t(\underset{\sim}{x}_n)) - \log \eta(t(\underset{\sim}{x}_n)] \geqq \log \phi(\hat{\theta}) - \log \phi(\eta)$$

$$\Leftrightarrow$$

$$\frac{1}{n} \log L(\hat{\theta}) \geqq \frac{1}{n} \log L(\eta) \quad . \quad \square$$

In the 'classical' case we have that $t(\underset{\sim}{x}_n) \in Z^d$ and all homomorphisms being of the form

$$(z_1,\ldots,z_d) \to \theta_1^{z_1},\ldots,\theta_d^{z_d} \ .$$

Then, if $t = (t_1,\ldots,t_d)$

$$\frac{1}{n} \log \hat{\theta}(t) - \frac{1}{n} \log \eta(t) = \frac{1}{n} \sum_{i=1}^{d} (\log \theta_i - \log \eta_i) t_i$$

and further

$$m_\theta(\eta) = E_\theta \sum_{i=1}^{d} (\log \theta_i - \log \eta_i) t_i(X) = \sum_{i=1}^{d} (\log \theta_i - \log \eta_i) E_\theta t_i(X) \quad ,$$

The system of equations (4.3) is equivalent to the system of equations for i=1,...,d

$$E_\theta t_i(X) = \frac{1}{n} t_i(\underset{\sim}{x}_n) \quad ,$$

showing that (4.3) is the generalisation of the classical result for exponential families about equating the value of the canonical statistic to its expectation.

In the classical case, Barndorff-Nielsen has shown that if $\mathcal{D}_F$ is open, then (4.3) has a solution.

The analogous condition in the general case is obtained as follows:

For $\theta, \eta \in \mathcal{D}_F$ and $\lambda \in \mathbb{R}$ we define the exponential function

$$\theta_\lambda(t) = \begin{cases} \theta^{1+\lambda}(t)\ \eta^{-\lambda}(t) & \text{if } t \in F \\ 0 & \text{otherwise} \end{cases}$$

Hölders inequality implies that if $\lambda \in [-1,0]$, $\theta_\lambda \in \mathcal{D}_F$.

We then consider:

4.6 Condition For all $\theta, \eta \in \mathcal{D}_F$ there is a $\lambda_0 > 0$ such that $\theta_{\lambda_0} \in \mathcal{D}_F$.

Again by Hölders inequality this implies $\theta_\lambda \in \mathcal{D}_F$ for $\lambda \in [-1,\lambda_0]$.

It says that in some sense θ is not on the "boundary" of $\mathcal{D}_F$, in fact in the classical case this would just mean relative openness of $\mathcal{D}_F$.

Before we show the result analogous to that of Barndorff-Nielsen we shall study the function

$$\psi_{\theta,\eta}(\lambda) = \log \phi(\theta_\lambda)$$

which is the analogue of the moment generating function in the classical case. We have

4.7 Lemma If condition 4.6 is satisfied $\psi_{\theta,\eta}$ is convex, finite and analytic on the interval $]-1,\lambda_0[$ and its n'th derivative is equal to

$$D^n\psi_{\theta,\eta}(\lambda) = E_{\theta_\lambda} u(t(X))^n$$

where $$u(t) = \log \frac{\theta(t)}{\eta(t)}.$$

Proof The proof is essentially identical to the classical case, see p. 8-9 of Johansen (1979). We shall give the arguments for the sake of completeness.

The convexity and finiteness is just Hölders inequality. Let $v(t) = \theta(t)/\eta(t)$. Then

$$\phi\left(\theta_{\alpha\lambda_1+(1-\alpha)\lambda_2}\right) = \phi(\theta)E_\theta v(t(X))^{\alpha\lambda_1+(1-\alpha)\lambda_2}$$

$$= \phi(\theta)E_\theta\left[\left(v(t(X))^{\lambda_1}\right)^\alpha \left(v(t(X))^{\lambda_2}\right)^{1-\alpha}\right]$$

$$\leqq \left(\phi(\theta)E_\theta v(t(X))^{\lambda_1}\right)^\alpha \left(\phi(\theta)E_\theta v(t(X))^{\lambda_2}\right)^{1-\alpha}$$

$$= \phi(\theta_{\lambda_1})^\alpha\cdot\phi(\theta_{\lambda_2})^{1-\alpha}.$$

If $\lambda \in]-1,\lambda_0[$ we have if $h \in \mathbb{R}$

$$\phi(\theta_{\lambda+h}) = \sum_x \theta_\lambda(t(x))v(t(x))^h \mu(x)$$

$$= \sum_x \sum_{n=o}^{\infty} \theta_\lambda(t(x))h^n u(t(x))^n \mu(x)/n!$$

To allow for interchangement of the summation signs we evaluate

$$\sum_{n=0}^{\infty} \frac{|hu(t(x))|^n}{n!} = e^{|hu(t(x))|} \leqq e^{hu(x)} + e^{-hu(x)}$$

whereby

$$\sum_{x} \sum_{n=0}^{\infty} |\theta_\lambda(t(x))h^n\, u(t(x))^n/n!\; \mu(x)|$$

$$\leqq \phi(\theta_{\lambda+h}) + \phi(\theta_{\lambda-h}) < \infty$$

if h is sufficiently small. Thus we get

$$\phi(\theta_{\lambda+h}) = \sum_{n=0}^{\infty} \frac{h^n}{n!} \sum_{x} \theta_\lambda(t(x))u(t(x))\mu(x)$$

$$= \phi(\theta_\lambda) \sum_{n=0}^{\infty} \frac{h^n}{n!} E\theta_\lambda u(t(X))^n$$

whereby the lemma follows. □

In particular we get

$$\psi'_{\theta,\eta}(0) = E_\theta\; u(t(X)) = m_\theta(\eta). \tag{4.4}$$

We are now ready to show

<u>4.8 Proposition</u> <u>If condition 4.6 is satisfied</u> (4.3) <u>has a solution</u>.

<u>Proof</u> By our previous result we know that there is a unique maximum likelihood estimate $\hat{\theta} \in \mathcal{D}_F$ with $F = F(t(x_n))$.

Let $\eta \in \mathcal{D}_F$ and define θ_λ as before with $\theta=\hat{\theta}$. The function

$$l(\lambda) = -\log L(\theta_\lambda) \quad , \quad \lambda \in \,]-1,\lambda_0[$$

has thus a local minimum for $\lambda = 0$ and since it is differentiable:

$$l(\lambda) = n\psi(\lambda) + \lambda \log\eta(t(\underset{\sim}{x}_n)) - (1+\lambda)\log\hat{\theta}(t(\underset{\sim}{x}_n))$$

we must have $l'(0) = 0$ or

$$n\psi'(0) = \log\hat{\theta}(t(\underset{\sim}{x}_n)) - \log\eta(t(\underset{\sim}{x}_n))$$

which gives

$$m_{\hat{\theta}}(\eta) = \frac{1}{n}[\log\hat{\theta}(t(\underset{\sim}{x}_n)) - \log\eta(t(\underset{\sim}{x}_n))]$$

which is (4.3). □

Note finally that <u>if</u> $t(\mathcal{X})$ <u>is finite, condition 4.6 is always satisfied</u> because $\mathcal{D}_F$ is equal to all the exponentials with positivity filter F.

5. Examples of general exponential families

The most basic example of an exponential family is the family of arbitrary distributions on a countable set $\mathcal{X}$.

5.1 Example Let $\mathcal{X}$ be a countable set and let ${}^{\mathcal{X}}\mathbb{N}_0$ denote the free commutative semigroup over $\mathcal{X}$ consisting of all maps

$$s : \mathcal{X} \to \mathbb{N}_0$$

with finite and non empty support (i.e. $\{x : s(x) \neq 0\}$ finite and non-empty) equipped with the composition of pointwise addition:

$$(st)(x) = s(x) + t(x) \quad .$$

Let $t : \mathcal{X} \to S = {}^{\mathcal{X}}\mathbb{N}_0$ be given as the canonical embedding:

$$t(x) = 1_{\{x\}} \quad .$$

Then, clearly

$$t_n(x_1,\ldots,x_n) = \sum_{i=1}^{n} 1_{\{x_i\}} = (n(x))_{x\in\mathcal{X}} \quad ,$$

where $n(x)$ is the number of j's with $x_j = x$. In other words, the statistic is simply the histogram.

S is graduated by the function

$$\tau(s) = \sum_{x\in\mathcal{X}} s(x)$$

and

$$\mathcal{Y}_n = t(\underset{\sim}{\mathcal{X}}_n) = \tau^{-1}(n) \quad .$$

The exponential functions on S are given by numbers $(p_x)_{x\in\mathcal{X}}$ with $p_x \geq 0$ and

$$\chi^{\underset{\sim}{p}}(s) = \prod_{x \in X} p_x^{s(x)} .$$

If we let $\|\underset{\sim}{p}\| = \sum_{x \in X} p_x$ we see that the exponential family corresponding to this situation is given by the family

$$\{P^{\underset{\sim}{p}} : p \in \mathrm{EXP}(S) \text{ and } \|\underset{\sim}{p}\| < \infty\}$$

as

$$P^{\underset{\sim}{p}}\{X_n = x\} = \frac{p_x}{\|\underset{\sim}{p}\|}$$

and $X_1, X_2, \ldots$ independent.

The filters of S are all of the form

$$F_A = \{s \in S : s(x) = 0 \text{ for } x \notin A\} .$$

where A is a subset of X, $A \neq X$.

The smallest filter containing an observation s_0 is given as

$$F_{s_0} = \{s \in S : s_0(x) = 0 \Rightarrow s(x) = 0\}$$

The maximum likelihood equation (4.3) reduces here to the 'classical' one:

$$E_{\underset{\sim}{p}} 1_{\{x\}} = \frac{n(x)}{n} \Leftrightarrow p_x = \frac{n(x)}{n} \quad \forall x \in X .$$

Note that the exchangeable σ-algebra exactly is the tail σ-algebra of the histograms. Thus, since this is trivial for any extreme point, i.e. any product of identical distributions, the discrete version of the Hewitt-Savage 0-1 law is a consequence of the results in this chapter, and our results also contain a version of de Finetti's (1931) theorem.

Similarly, we can obtain families as the set of all symmetric distributions on $\mathbb{Z}$ by considering the statistic

$$t(x) = 1_{\{|x|\}} \quad t : \mathbb{Z} \to {}^{\mathbb{N}_0}\mathbb{N}_0 .$$

<u>5.2 Example</u> Let $S = (\mathbb{N}, \vee)$ and let $X = S$ and $t(x) = x$.

All exponentials then satisfy

$$\chi(x \vee x) = \chi(x) = \chi(x)^2$$

whereby χ is an indicator function of a set F. But 1_F is an exponential function if and only if F is a filter and the filters of S are all the <u>intervals</u> $[1,\theta] = \{1,\ldots,\theta\}$, $\theta \in \mathbb{N} \cup \{\infty\}$.

But $\phi(\theta) = |[1,\theta]| = \theta$ such that the corresponding exponential family is the family of <u>uniform distributions</u>

$$P^{\theta}\{X = x\} = \frac{1}{\theta}\, 1_{[1,\theta]}(x), \quad \theta \in \mathbb{N} .$$

If we let $t_0 = x_1 \vee \ldots \vee x_n$, $F_{t_0} = [1,t_0]$ such that the maximum likelihood estimate of θ by Proposition 4.4 is

$$\hat{\theta} = t_0 = x_1 \vee \ldots \vee x_n = x_{(n)} .$$

Combining this example with the classical case we can see that standard exponential families with unknown truncation points etc. are general exponential families.

<u>5.3 Example</u> Let $X = \mathbb{N}$ and let S be the semigroup of positive integers equipped with <u>multiplication</u> and let $t(x) = x$. $(\mathbb{N},\cdot)$ is isomorphic to the free commutative semigroup <u>with unit</u> over the prime numbers (1 not counted as a prime number).

If we namely let $\mathbb{P}$ denote the prime numbers different from 1,

$$\hat{\theta}_p = \frac{n_p(x_1 \dots x_n)}{n + n_p(x_1 \dots x_n)} .$$

<u>5.4 Example</u> (Martin-Löf's model for item analysis.)

We consider again the problem in example 9.7 of chapter II.

X is the set of subsets of a finite set M of <u>items</u> or <u>questions</u>. X_i is the set of questions that the i'th person, $i = 1,\dots,n$, $n \in \mathbb{N}$ answers right.

Martin-Löf (1970) proposed the model having X_i independently and identically distributed as

$$P\{X_i = x\} = \frac{\prod_{\mu \in x} \alpha_\mu}{\gamma_{|x|}(\alpha_\mu, \mu \in M)} \; p(|x|) \tag{5.2}$$

where $p:\{0,\dots,m\} \to [0,1]$ satisfies $\sum_{s=0}^{m} p(s) = 1$ and $0 < \alpha_\mu < \infty \quad \forall \mu$.

Obviously, this model is obtained from the conditional Rasch model described in chapter II by letting $|X_i|$ have an arbitrary distribution:

$$p(s) = P\{|X_i| = s\} .$$

We shall here let our semigroup be a subsemigroup of the product of the free semigroups over $\{0,1,\dots,m\}$ and M, namely the subsemigroup generated by the values of the statistic

$$t : X \to {}^{\{0,1,\dots,m\}}\mathbb{N}_0 \times {}^{M}\mathbb{N}_0$$

given as

$$t(x) = (1_{\{|x|\}}, 1_x) .$$

the correspondence

$$x \leftrightarrow (n_p(x))_{p\in \mathbb{P}}$$

where

$$x = \prod_{p\in \mathbb{P}} p^{n_p(x)}$$

is the prime factorisation of x, i.e.

$$n_p(x) = \sup\{\nu : p^\nu | x\} ,$$

clearly establishes an isomorphism between $(\mathbb{N},\cdot)$ and $({}^{\mathbb{P}}\mathbb{N})^1$.

It follows that the corresponding exponential family is given by

$$P_\theta\{X = x\} = \frac{\prod_{p\in \mathbb{P}} \theta_p^{n_p(x)}}{\phi(\theta)} \tag{5.1}$$

where $(\theta_p)_{p\in \mathbb{P}}$ is a family of non-negative real numbers.

We easily get from (5.1) that the numbers $(n_p(X))_{p\in \mathbb{P}}$ are independent and geometrically distributed as

$$P_\theta\{n_p(X) = \nu\} = (1-\theta_p)\theta_p^\nu \quad \nu = 0,1,\ldots$$

whereby

$$\phi(\theta)^{-1} = \prod_{p\in \mathbb{P}} (1-\theta_p)$$

and

$$\phi(\theta) < \infty \Leftrightarrow \sum_{p\in \mathbb{P}} \theta_p < \infty .$$

The maximum likelihood estimate $\hat{\theta}_p$ based on observations $X_1 = x_1,\ldots,X_n = x_n$ is given as

This semigroup is obviously graduated by

$$\phi(f,g) = \sum_{s=0}^{m} f(x)$$

such that we just have to find all its exponential functions. These are again defined by their values of the generators, so let us define for $\xi \in \mathrm{EXP}(S)$

$$\psi(x) = \xi(t(x))$$

and note that ψ corresponds to an exponential function if and only if

$$\psi(x \cup \{\lambda\})\psi(y \cup \{\mu\}) = \psi(x \cup \{\mu\})\psi(y \cup \{\lambda\}) \tag{5.3}$$

for all $x,y \subseteq M \setminus \{\lambda,\mu\}$, $\lambda \in M$, $\mu \in M$.

If it so happens that ψ is everywhere positive, one can easily deduce by an argument completely analogous to the one used to show Proposition II 9.19, that ψ must be of the form

$$\psi(x) = \prod_{\mu \in x} \alpha_\mu \cdot p(|x|) , \tag{5.4}$$

which again implies that (5.2) holds for the distribution of X w.r.t. P_ψ.

If this is not the case, the situation can be slightly more complicated. Of course, the probabilities are then degenerate, but we shall in the sequel find all those degenerate models and it will, hopefully, then be clear that these in a natural way belong to the item analysis problem.

First let

$$T^+ = \{s \mid \exists x : \psi(x) > 0 \text{ and } |x| = s\} .$$

Define further the sets for $s \in T^+$

$$A_s^0 = \{\lambda \in M : \psi(a \cup \lambda) = 0 \quad \forall\ a \subseteq M \setminus \{\lambda\} \text{ with } |a| = s-1\}$$

$$A_s^1 = \{\lambda \in M : \psi(a) = 0 \quad \forall\ a \subseteq M \setminus \{\lambda\} \text{ with } |a| = s\}$$

$$A_s^+ = \{\lambda \in M : \exists a, b \in M \setminus \{\lambda\}, |a| = s, |b| = s-1 : \psi(a)\psi(b \cup \lambda) > 0\}$$

and note that A_s^0, A_s^+, A_s^1 constitute a disjoint partitioning of M.

We first show

<u>5.5 Lemma</u> <u>The following relations hold for any</u> $s', s'' \in T^+$:

(i) $$(A_{s''}^0 \subseteq A_{s'}^0) \vee (A_{s''}^0 \supseteq A_{s'}^0 \cup A_{s'}^+)$$

(ii) $$(A_{s''}^1 \subseteq A_{s'}^1) \vee (A_{s''}^1 \supseteq A_{s'}^+ \cup A_{s'}^1)$$

<u>Proof</u> We shall first show (i) and this is done the following way.

We assume that there exists a $\lambda \in A_{s''}^0 \setminus A_{s'}^0$ and will then show that $A_{s''}^0 \supseteq A_{s'}^0 \cup A_{s'}^+$. This is done by assuming the existence of a $\mu \notin A_{s''}^0$ and showing that μ then cannot be an element neither of $A_{s'}^0$ nor of $A_{s'}^+$.

Since $\lambda \in A_{s''}^0$ and $\mu \notin A_{s''}^0$ we must have $\lambda \neq \mu$ and there must be a set $a \subseteq M \setminus \{\lambda, \mu\}$ with $|a| = s'' - 1$ and

$$\psi(a \cup \{\mu\}) > 0 .$$

For any $b \subseteq M \setminus \{\lambda, \mu\}$ we have as usual

$$\psi(b \cup \{\lambda\})\psi(a \cup \{\mu\}) = \psi(a \cup \{\lambda\})\psi(b \cup \{\mu\}) \quad .$$

Since $\lambda \in A^0_{s''}$, $\psi(a \cup \{\lambda\}) = 0$, implying

$$\psi(b \cup \{\lambda\}) = 0 \text{ for all } b \subseteq M \setminus \{\lambda,\mu\}. \tag{5.5}$$

But, since $\lambda \notin A^0_{s'}$ there must be a set containing λ with positive ψ and such a set must then contain μ as well, i.e., there is a $b^* \subseteq M \setminus \{\lambda,\mu\}$ such that

$$\psi(b^* \cup \{\lambda,\mu\}) > 0$$

and $|b^*| = s' - 2$. But this means that μ is <u>not</u> an element of $A^0_{s'}$. It remains to be shown that it cannot be an element of $A^+_{s'}$ either. We then assume $\mu \in A^+_{s'}$ and will show that this leads to a contradiction. If $\mu \in A^+_{s'}$ there must be a set $b^{**} \subseteq M \setminus \{\mu\}$ such that

$$\psi(b^{**}) > 0 \quad , \quad |b^{**}| = s'$$

but, since also (5.5) holds, we must have $b^{**} \subseteq M \setminus \{\lambda,\mu\}$. Since $|b^{**}| = s'$ and $|b^*| = s' - 2$, there must be a $\kappa \in b^{**} \setminus b^*$ and we have the relation

$$0 < \psi(b^* \cup \{\lambda,\mu\})\psi(b^{**}) = \psi(b^* \cup \{\mu,\kappa\})\psi((b^{**} \setminus \{\kappa\}) \cup \{\lambda\})$$

But then $\psi((b^{**} \setminus \{\kappa\}) \cup \{\lambda\}) > 0$ which contradicts (5.5). Thus (i) is proved.

(ii) follows from (i) by symmetry : the operation

$$\psi \to \tilde{\psi}$$

where

$$\tilde{\psi}(x) = \psi(x^c)$$

preserves the exponential property of ψ and makes

$$A_s^+ \to A_{m-s}^+$$

$$A_s^0 \to A_{m-s}^1$$

$$A_s^1 \to A_{m-s}^0 .$$

This completes the proof of the lemma. □

We can then show further:

<u>5.6 Lemma</u> <u>If</u> $s' < s''$ <u>then</u>

<u>either</u> (a) $A_{s'}^0 = A_{s''}^0 , \quad A_{s'}^+ = A_{s''}^+ , \quad A_{s'}^1 = A_{s''}^1$

(i.e. the partitions are identical)

<u>or</u> (b) $A_{s''}^1 \supseteq A_{s'}^+ \cup A_{s'}^1$

<u>Proof</u> We have obviously one of the following relations

$$A_{s''}^0 \not\subseteq A_{s'}^0 , \quad A_{s''}^0 \not\supseteq A_{s'}^0 \quad \text{or} \quad A_{s''}^0 = A_{s'}^0 .$$

The first case cannot happen since then, by Lemma 5.5, (i)

$$A_{s''}^0 \supseteq A_{s'}^0 \cup A_{s'}^+$$

which implies $s'' \leq s'$.

If we have the second relation, Lemma 5.5 (i) gives

$$A_{s'}^0 \supseteq A_{s''}^0 \cup A_{s''}^+$$

which gives (b) by taking complements.

In the third case, if also $A^1_{s''} = A^1_{s'}$ (a) holds. If not, we have either

$$A^1_{s'} \not\subseteq A^1_{s''} \text{ or } A^1_{s'} \not\supseteq A^1_{s''} .$$

In the first case, we get by Lemma 5.5 (ii) that

$$A^1_{s'} \supseteq A^+_{s''} \cup A^1_{s''}$$

which implies $s' \geqq s''$ so that this cannot happen.

In the second case, Lemma 5.5 (ii) gives

$$A^1_{s''} \supseteq A^+_{s'} \cup A^1_{s'} ,$$

which is (b). □

Further we have

5.7 Corollary *Any pair $A^+_{s'}, A^+_{s''}$ are either identical or disjoint.*

It follows now by an argument analogous to that used in the proof of Proposition II 9.19 that there exist positive numbers

$$\alpha_\mu, \mu \in \bigcup_{s \in T^+} A^+_s \text{ , and } p(s),\ s \in T^+$$

such that

$$\psi(x) = p(|x|) \cdot \prod_{\mu \in A^+_{|x|}} \alpha_\mu \tag{5.6}$$

$$\text{if } |x| \in T^+,\ x \supseteq A^1_{|x|}\ \ x \cap A^0_{|x|} = \emptyset ,$$

and otherwise equal to zero.

Also, the set T^+ can be partitioned into sets $T^+_{(1)}, \ldots, T^+_{(k)}$ such that (a) of Lemma 5.6 holds for $s', s'' \in T_{(i)}$ whereas (b) holds for $s' \in T_{(i)}, s'' \in T_{(j)}$ with $i < j$.

Similarly the set of questions can be partitioned into groups

$$M_1 = A_s^+ \cup A_s' \quad \text{for } s \in M_1$$

$$M_{i+1} = (A_s^+ \cup A_s^1) \setminus \bigcup_{j=1}^{i-1} M_j \quad \text{for } s \in T_{(i+1)}$$

$$M_{k+1} = A_s^0 \quad \text{for } s \in T_{(k)}$$

of questions that relative to each other are incomparably difficult in the sense that any person answering a question in group M_{i+1} will automatically have answered all questions in groups $M_j, j \leqq i$.

6. Bibliographical notes

The semigroup theory is basically taken from Clifford and Preston (1961) and Petrich (1977), apart from the characterisation of strongly separative semigroups.

An excellent treatment of harmonic analysis on such semigroups (i.e. the theory of Laplace transforms) has now appeared in Berg, Christensen and Ressel (1984).

The general exponential families introduced here were introduced by the author (1975) and all results in this chapter apart from the estimating equation (Proposition 4.5) which is new, Martin-Löf's local limit theorem (Martin-Löf (1970)) and its corollaries, and the examples are contained in that paper.

The proof of the fact that all extreme points correspond to exponential functions is based on a proof by Neveu (1964) of the fact that all harmonic functions for a random walk on the integer lattice are mixtures of exponentials.

A more general theorem has now been shown by Ressel (1985), also describing an elegant connection to the theory of positive definite functions.

Martin-Löf's model for the item analysis problem was discussed by Martin-Löf (1974) but without finding all the extreme points.

Johansen (1967) also characterised one-dimensional exponential families as extreme points of the set of distributions with $t(x_1)+\ldots+t(x_n)$ sufficient for all $n \in \mathbb{N}$. See also a related result of Andersen (1970).

Barndorff-Nielsen (1970,1978) and Chentsov (1972) have considered so called <u>extended</u> exponential families.

The extended families are related to the families considered here and in 'the classical case' discussed in section 3, they are identical to those here considered, see Martin-Löf (1974).

The essential difference (apart from allowing general semigroups) between the extended and general exponential families is that the extended ones are defined via an embedding of the semigroup S into the semigroup $(\mathbb{R}^k, +)$ for some k. For details of this see Lauritzen (1975).

Generalisations of the notion of an exponential family have e.g. also been considered by Johansen (1977), based on semigroup considerations as well.

The general exponential families also occur as solutions to certain maximal entropy problems, see Lauritzen (1984) for a discussion of this aspect.

An unusual, but close, connection between our results and probabilistic logic can be found in Fenstad (1967, 1968).

Chapter IV

GENERAL REPETITIVE STRUCTURES OF POLISH SPACES. PROJECTIVE STATISTICAL FIELDS

0. Probability measures on Polish spaces

We shall here without proofs give a list of some basic properties of probability measures on Polish spaces. For details, the reader is referred to Billingsley (1968) or Asmussen (1978).

A topological space X is Polish if it is separable (countable base for open sets) metrisable (there is a metric defining the topology) and topologically complete (there is a metric, defining the topology such that the metric space is complete).

A subspace X of a Polish space $\hat{X}$ is Polish if and only if it is a G_δ - set, i.e. there are open sets $(O_n)_{n\in\mathbb{N}}$ such that $X = \bigcap_n O_n$.

A compact space is Polish if and only if it is metrisable which is the case if and only if it is separable.

For X being a Polish space we let $C(X)$ denote the space of real-valued, bounded continuous functions of X.

We let further $\mathbb{P}(X)$ denote the space of Borel-probability measures, in the following just denoted probability measures.

(The Borel σ-algebra is here the smallest σ-algebra generated by the open sets).

The weak topology on $\mathbb{P}(X)$ is the topology induced by the maps

$$P \to \int f dP \quad , \quad f \in C(X) \tag{0.1}$$

and we write

$P_n \overset{w}{\to} P$ if and only if $\int f dP_n \to \int f dP$ for all $f \in C(X)$.

If $C(X)$ is endowed with the uniform topology, the maps

$$f \to \int f dP \quad , \quad P \in \mathbb{P}(X) \tag{0.2}$$

are all positive continuous linear functionals on $C(X)$ and, in fact, there is a one-to-one correspondence between $\mathbb{P}(X)$ and the set of positive continuous linear functionals on $C(X)$ that are also <u>order continuous</u> and have value 1 on the constant 1_X:

$$f_n \searrow 0 \Rightarrow P(f_n) \to 0 \tag{0.3}$$

$$P(1_X) = 1$$

the correspondence being given as

$$[f \to P(f)] \leftrightarrow [f \to \int f dP]$$

or, in terms of sets:

$$P(F) = \inf_{f \in C(X),\ f \geqq 1_F} P(f), \text{ F closed.}$$

If X is a <u>compact</u> Polish space, (0.3) is automatically fulfilled for any continuous positive linear functional (Radon measure) on $C(X)$.

Further, there is a countable class $\mathcal{D} \subseteqq C(X)$ defining the weak topology, i.e. such that

$$P_n \overset{w}{\to} P \Leftrightarrow \int f dP_n \to \int f dP \ \forall \ f \in \mathcal{D} \ .$$

$\mathbb{P}(X)$, endowed with the weak topology <u>is itself a Polish space and compact if and only if</u> X <u>is compact</u>.

If $O \subseteqq X$ is <u>open</u>, the function

$$P \to P(O)$$

is <u>lower semicontinuous</u> such that the subsets of $\mathbb{P}(X)$

$$\{P : P(O) > t\}$$

are <u>open</u> for all $t \in \mathbb{R}$.

There exists also a countable class $\mathcal{D} \subseteq C(X)$ separating points and closed sets giving rise to the following <u>compactification</u> of X.

For $f \in \mathcal{D}$, let

$$a_f = \inf_{x \in X} f(x)$$

$$b_f = \sup_{x \in X} f(x)$$

and define the embedding:

$$i : X \to \prod_{f \in \mathcal{D}} [a_f, b_f]$$

$$x \overset{i}{\to} (f(x))_{f \in \mathcal{D}} \,.$$

If $\hat{X}$ denotes the closure of $i(X)$, $\hat{X}$ is a compactification of X making $\hat{X}$ a compact Polish space and thus X (or rather $i(X)$) a G_δ - subset of $\hat{X}$.

This compactification has further the following properties.

There is a unique correspondence between $\mathbb{P}(X)$ (in fact a homeomorphism) and those elements of $\mathbb{P}(\hat{X})$ concentrated on $i(X)$, making $\mathbb{P}(X)$ a G_δ - subset of $\mathbb{P}(\hat{X})$.

There is a unique correspondence between functions in $C(\hat{X})$ and the functions in $C(X)$ that can be continuously extended to $\hat{X}$. The latter class is the smallest closed vector sublattice of $C(X)$ containing $\mathcal{D}$ and the constant 1_X.

If X and Y are Polish spaces, a <u>Markov kernel</u> Π from X to Y is a continuous map from X to $\mathbb{P}(Y)$. We shall write

$$X \overset{\Pi}{\to} Y$$

Such a kernel can be given in several ways, e.g. as a 'genuine' kernel

$$\Pi : \mathcal{X} \times \mathcal{B}(\mathcal{Y}) \to [0,1]$$

where $\mathcal{B}(\mathcal{Y})$ is the Borel subsets of $\mathcal{Y}$ satisfying:

i) $$\Pi(x,\cdot) \in \mathbb{P}(\mathcal{Y}) \text{ for all } x \in \mathcal{X}$$

ii) $$x \to \int_{\mathcal{Y}} f(y)\Pi(x,dy) \in C(\mathcal{X}) \text{ for all } f \in C(\mathcal{Y}) .$$

A Markov kernel Π induces continuous maps

$$\Pi : \mathbb{P}(\mathcal{X}) \to \mathbb{P}(\mathcal{Y})$$

where

$$(\Pi P)(O) = \int_{\mathcal{X}} \Pi(x,O)P(dx)$$

and

$$\Pi^* : C(\mathcal{Y}) \to C(\mathcal{X})$$

$$(\Pi^*f)(x) = \int_{\mathcal{Y}} f(y)\Pi(x,dy) .$$

As hinted by our notation, we shall not distinguish between the map $\Pi : \mathcal{X} \to \mathbb{P}(\mathcal{Y})$, the map $\Pi : \mathbb{P}(\mathcal{X}) \to \mathbb{P}(\mathcal{Y})$ and the kernel Π.

Π^* is the <u>adjoint</u> of Π: Let

$$<f,P> = \int f dP$$

then $\forall f \in C(\mathcal{Y})$, $P \in \mathbb{P}(\mathcal{X})$

$$<f,\Pi P> = <\Pi^* f,P>.$$

Suppose now that we have a Markov kernel $\Pi : \mathcal{X} \to \mathcal{Y}$. We want to construct compactifications $\hat{\mathcal{X}}$ and $\hat{\mathcal{Y}}$ of $\mathcal{X}$ and $\mathcal{Y}$ in such a way that $\hat{\mathcal{X}}$ and $\hat{\mathcal{Y}}$ are Polish and Π in a unique way can be extended to a Markov kernel $\hat{\Pi}$ from $\hat{X}$ to $\hat{\mathcal{Y}}$.

Choose then first a countable class $\mathcal{D} \subseteqq C(\mathcal{Y})$ separating points and closed sets and let $\hat{\mathcal{Y}}$ be the compactification described earlier, making it possible to identify $C(\hat{\mathcal{Y}})$ with the smallest closed vector sublattice of $C(\mathcal{Y})$ containing $\mathcal{D}$ and the constant function.

Choose then $\tilde{\mathcal{D}}$ as a dense countable set in this sublattice.

Choose now $E \subseteqq C(X)$ to be countable, to separate points and closed sets and such that

$$E \supseteqq \Pi^*(\tilde{\mathcal{D}}).$$

Let $\hat{X}$ be the now familiar compactification made by the countable class E.

Define a linear map

$$\hat{\Pi}^* : C(\hat{\mathcal{Y}}) \to C(\hat{X})$$

as follows.

Firstly, for any $f \in \tilde{\mathcal{D}}$ let $\hat{f}$ denote its extension to $\hat{\mathcal{Y}}$ and let

$$\hat{\Pi}^*(\hat{f}) = \widehat{\Pi^*(f)}$$

in other words, the extension $\hat{\Pi}^*$ maps the extension of f into the extension of its image by Π^*.

Since $\{\hat{f},\ f \in \tilde{\mathcal{D}}\}$ is dense in $C(\hat{\mathcal{Y}})$ this defines a unique linear, positive, continuous map from $C(\hat{\mathcal{Y}})$ to $C(\hat{X})$ mapping the identity into itself.

$\hat{\Pi}$ is now well defined as the adjoint to $\hat{\Pi}^*$, i.e. for any $x \in X$

$$\hat{\Pi}(x,\cdot)$$

is the unique probability measure on $\hat{\mathcal{Y}}$ satisfying

$$\int f(y)\hat{\Pi}(x,dy) = (\hat{\Pi}^{*}(f))(x)$$

for all $f \in C(\hat{\mathcal{Y}})$.

We shall be interested in compositions of two Markov kernels Π and R

$$\mathcal{X} \overset{\Pi}{\rightarrow} \mathcal{Y} \qquad \mathcal{Y} \overset{R}{\rightarrow} \mathcal{Z}$$

defined simply as $R\Pi$, the corresponding compositions of the maps from the corresponding spaces of probabilities.

If $t : \mathcal{X} \rightarrow \mathcal{Y}$ is a continuous map it induces a Markov kernel Π^t as

$$\Pi^t(\varepsilon_x) = \varepsilon_{t(x)}$$

where ε_x is one point measure in x.

For such Markov kernels we obviously have

$$\Pi^{s o t} = \Pi^s \Pi^t.$$

The Markov kernel so induced by the identical map $\mathcal{X}$ is denoted $I_{\mathcal{X}}$ and called the identity.

If t is a map $\mathcal{X} \rightarrow \mathcal{Y}$ such that Π^t has an inverse Markov kernel $Q : \mathcal{Y} \rightarrow \mathcal{X}$

$$\Pi^t Q = I_{\mathcal{Y}},$$

Q is the <u>conditional</u> distribution of X given t(X) = y for all measures $P \in \mathbb{P}(\mathcal{X})$ satisfying

$$Q\Pi^t P = P.$$

We finally recall from Choquet's theorem that the extreme points of a compact, metrisable convex subset of a linear, locally convex topological space is a G_δ- set and thus Polish (see e.g. Phelps (1966)).

1. Projective systems of Polish spaces and Markov kernels

Consider a set I with a partial order $<$ such that I is directed to the right:

$$\forall i,j \quad \exists k \in I : i < k,\ j < k.$$

We say that I is countable at infinity if I contains a cofinal sequence, i.e. a sequence $(i_n)_{n \in \mathbb{N}}$ of elements in I such that

$$\forall i \in I \ \exists n \in \mathbb{N} : i < i_n.$$

In the rest of this chapter we shall always, unless the opposite is explicitly stated, assume that I is countable at infinity.

A projective system of Polish spaces and Markov kernels (with base I) consists of a family

$$(X_i)_{i \in I}$$

of Polish spaces, and a family of Markov kernels

$$P_{ij} : X_j \to X_i \qquad i < j$$

satisfying the 'Chapman-Kolmogorov equations':

$$\forall\, i < j < k : P_{ij} P_{jk} = P_{ik}.$$

These Markov kernels shall be denoted projections.

Let M be the convex set of consistent families of probability measures, i.e. families

$$\mu = (\mu_i)_{i \in I}$$

satisfying

$$\mu_i \in \mathbb{P}(X_i)$$

and

$$P_{ij}\mu_j = \mu_i \quad \text{for all } i < j.$$

Let E denote the extreme points of M and endow M and E with the topology induced by all the maps

$$\mu \to \int f_i d\mu_i \ , \ f_i \in C(X_i),$$

i.e. the product of the weak topologies on $\mathbb{P}(X_i)$, restricted to M (and E).

We then have the following integral representation theorem:

<u>1.1. Proposition</u> <u>E and M are Polish</u>. <u>There is a one-to-one affine correspondence between M and $\mathbb{P}(E)$</u>

$$\mu \leftrightarrow P_\mu$$

<u>given as</u>

$$\int_{X_i} f_i d\mu_i = \int_E \left(\int_{X_i} f_i de_i\right) dP_\mu(e) \ .$$

<u>Proof</u> Assume first that $I = \mathbb{N}$ with the usual ordering and that X_n are all compact. The result will then follow if we can show that M is a compact metrisable Choquet simplex. Clearly the only non-trivial statement here is the simplex condition.

So let $\mu = (\mu_n)_{n\in\mathbb{N}}$ and $\nu = (\nu_n)_{n\in\mathbb{N}}$ be consistent families of bounded positive measures. We have to show that there is a consistent family which is the least upper bound of μ and ν, in the pointwise ordering.

Define now

$$\gamma_n = \lim_{N\to\infty} P_{nN}(\mu_N \vee \nu_N) = \lim_{N\to\infty} \gamma_n^N$$

We shall show that this limit is well defined and that $\gamma = (\gamma_n)_{n\in N}$ is the upper bound that we are looking for.

But

$$(\mu_n \vee \nu_n) \leqq P_{nN}(\mu_N \vee \nu_N) \leqq P_{nN}P_{N,N+1}(\mu_{N+1} \vee \nu_{N+1})$$

$$= P_{nN+1}(\mu_{N+1} \vee \nu_{N+1}) \leqq \gamma_n \qquad (1.1)$$

which implies that the sequence $(\gamma_n^N)_{N\in\mathbb{N}}$ is increasing. Since

$$\gamma_n^N(1_{X_n}) = (\mu_N \vee \nu_N)(1_{X_N}) \leqq \mu_N(1_{X_N}) + \nu_N(1_{X_N})$$

$$= \mu_n(1_{X_n}) + \nu_n(1_{X_n})$$

it is also norm bounded and therefore convergent. Finally (γ_n) is consistent since

$$P_{mn}\gamma_n = P_{mn} \lim \gamma_n^N = \lim \gamma_m^N = \gamma_m.$$

(1.1) now gives that $\gamma \geqq \mu,\nu$.

But if $\gamma' \geqq \mu,\nu$, we have for all $n<N$

$$\gamma_n' = P_{nN}(\gamma_N') \geqq P_{nN}(\mu_N \vee \nu_N) \to \gamma_n.$$

And thus that $\gamma' \geqq \gamma$. Then M is a Choquet simplex and the result follows.

If the spaces are <u>not</u> compact we proceed as follows (by compactification).

Choose countable classes $\mathcal{D}_n \subseteqq C(X_n)$ recursively as follows:

$\mathcal{D}_1$ is a countable subclass of $C(X_n)$ separating points and closed sets.

$\tilde{\mathcal{D}}_n$ is a countable class, dense in the smallest closed linear vector-sublattice of $C(X_n)$ containing $\mathcal{D}_n$ and the constant 1_{X_n}.

$\mathcal{D}_{n+1}$ is a countable class, separating points and closed sets and having

$$\mathcal{D}_{n+1} \supseteq P^*_{n,n+1}(\tilde{\mathcal{D}}_n).$$

The classes $\mathcal{D}_n$ now define compactifications $\hat{X}_n$ of X_n such that the family (P_{mn}) of Markov kernels in a unique way can be extended to $\hat{P}_{mn}$.

Letting $\hat{M}$ be the consistent families in the projective system

$$(\hat{X}_n)_{n\in\mathbb{N}}, \ (\hat{P}_{mn}, \quad m < n)$$

and $\hat{E}$ the corresponding set of extreme points, we get from the compact case just considered that any $\hat{\mu} \in \hat{M}$ can be uniquely represented by $P_{\hat{\mu}}$ being a probability on $\hat{E}$.

But M can then be identified with the face F of $\hat{M}$ of families $\hat{\mu}$ with

$$\hat{\mu}_n(X_n) = 1, \ n \in \mathbb{N}$$

But F is a G_δ - set since if $X_n = \bigcap_{p\in\mathbb{N}} O_{np}$

$$F = \bigcap_n \bigcap_p \bigcap_k \{\hat{\mu} \in \hat{M} : \hat{\mu}_n(O_{np}) > 1 - \tfrac{1}{k}\}$$

and these are all open. Thus M is Polish and

$$E = \hat{E} \cap F$$

is Polish as well. Finally, if $\mu \in F$, P_μ must be concentrated on E since

$$1 = \mu_n(O_{np}) = \int_{\hat{E}} e_n(O_{np})\,dP_\mu(e)$$

by letting $f_\nu \nearrow 1_{O_{np}}$ pointwise, implying

$$e_n(O_{np}) = 1 \text{ almost surely } P_\mu$$

and thus

$$e_n(X_n) = e_n(\bigcap_p O_{np}) = 1 \quad \text{a.s. } P_\mu \,.$$

Finally, the case $I \neq \mathbb{N}$ is solved by noticing that any family, consistent in the reduced projective system

$$(X_{i_n})_{n \in \mathbb{N}},\ P_{i_m i_n},\ m < n,$$

where $(i_n)_{n \in \mathbb{N}}$ is a cofinal sequence, in a unique way can be extended to the entire system by letting

$$\mu_i = P_{i i_n} \mu_{i_n} \quad \text{for all } i_n > i. \qquad \square$$

Note that if X_i <u>are all compact</u>, E <u>is non empty</u>, whereas E might be empty in other cases. (Then M is empty).

E, together with the Markov kernels

$$P_i : E \to X_i$$

$$P_i e = e_i \,.$$

(that shall be called <u>the canonical projections</u>), <u>represents the projective limit</u> of the system and we write

$$E = \varprojlim_{i \in I} X_i$$

in the sense that if $(\mu_i)_{i \in I}$ is any consistent family of <u>Markov</u>

<u>kernels</u>

$$\mu_i : \mathcal{Y} \to X_i, \ i \in I$$

from a Polish space $\mathcal{Y}$ to X_i, i.e.

$$P_{ij}\mu_j = \mu_i,$$

then there exists a unique (not necessarily continuous) Markov kernel $\mu : \mathcal{Y} \to E$ such that

$$\mu_i = P_i\mu.$$

μ is simply constructed by letting for each $y \in \mathcal{Y}$

$$\mu(y,\cdot)$$

be the representing measure of $(\mu_i(y,\cdot))_{i\in I}$.

The term 'projective limit' has not been used in the strict sense of category theory. It is <u>not</u> true that the category of Polish spaces and Markov kernels is closed under projective limits along directed index sets containing a cofinal sequence. If the set of consistent families is sufficiently complicated, there might exist spaces $\mathcal{Y}$ and kernels μ_i such that the Markov kernel μ defined above is <u>not</u> continuous.

The 'reason' is that the map between the probabilities on the extreme points of a convex set and the set itself, given by the Choquet representation, is <u>not</u> always continuous both ways but sometimes only in the direction $P_\mu \to \mu$, from the probability to its barycentre. In fact, Proposition 9.10 of Phelps (1966) gives that it is continuous if and only if the simplex is a Bauer simplex (i.e. the extreme points are closed) but always measurable. We shall however use the term and the notation for convenience but also to support the mind when thinking about our problems. Note that none of our results in the following are derived as consequences of such a continuity.

In the case where all the Markov kernels are induced by coherent continuous maps

$$p_{ij} : X_j \to X_i \quad p_{ij}p_{jk} = p_{ik} \text{ for } i < k.$$

We get as a corollary the following well known version of Kolmogorov's consistency theorem:

<u>1.2 Corollary</u> <u>To any consistent family</u> $(\mu_i)_{i\in I}$ <u>of probability measures</u>:

$$\mu_i(B_i) = \mu_j(p_{ij}^{-1}(B_i)) \quad i < j$$

<u>there is a unique probability</u> μ <u>on</u>

$$\lim_{\substack{\longleftarrow \\ i\in I}} X_i = \{(x_i)_{i\in I} : p_{ij}(x_j) = x_i\}$$

<u>having</u> μ_i <u>as marginals</u>

$$\mu_i(B_i) = \mu(p_i^{-1}(B_i)),$$

<u>where</u>

$$p_i(x_j)_{j\in I} = x_i.$$

<u>Proof</u> One just has to verify that the extreme consistent families are the consistent families of one point measures. □

2. Projective statistical fields

A projective statistical field is defined to consist of a projective system

$$(\mathcal{X}_i)_{i\in I},\ (P_{ij})_{i<j}$$

of Polish spaces and Markov kernels, a (Polish) parameter space Θ and a consistent family

$$\mu_i : \Theta \to \mathcal{X}_i,\ i \in I$$

$$P_{ij}\mu_j = \mu_i$$

of Markov kernels, and we write

$$\mathcal{F} = \{(\mathcal{X}_i)_{i\in I},\ (P_{ij})_{i<j},\ \Theta\ ,\ (\mu_i)_{i\in I}\}$$

We think of this as a family of 'extending' sample spaces $(\mathcal{X}_i)_{i\in I}$ and for each of these we have a 'usual' statistical model, namely the one that the random element X_i taking values in $\mathcal{X}_i$ has the distribution $\mu_i(\theta,\cdot)$ where $\theta \in \Theta$ is unknown.

In chapter II we considered the situation where $I = \mathbb{N}$ and the projective system in question was

$$\underset{\sim}{\mathcal{X}}_n = \mathcal{X}_1 \times \ldots \times \mathcal{X}_n$$

with projections induced by the coordinate projection mappings

$$p_{mn} : \underset{\sim}{\mathcal{X}}_n \to \underset{\sim}{\mathcal{X}}_m \qquad m < n$$

$$p_{mn}(x_1,\ldots,x_n) = (x_1,\ldots,x_m).$$

We discussed various notions of sufficiency and we see that the notion of a sufficient and transitive sequence has an immediate generalisation in the above given framework. Let now

$$t_i : \mathcal{X}_i \to \mathcal{Y}_i, \quad i \in I$$

be a family of continuous surjective maps into Polish spaces $(\mathcal{Y}_i)_{i \in I}$.

In analogy with Propositions I 1.3, I 1.4 and II 2.4 we say that $(t_i)_{i \in I}$ is <u>sufficient</u> in the field $\mathcal{F}$ if there exists Markov kernels

$$Q_i : \mathcal{Y}_i \to \mathcal{X}_i$$

and for $i < j$

$$Q_{ij} : \mathcal{Y}_j \to \mathcal{Y}_i$$

with

$$Q_{ij} Q_{jk} = Q_{ik} \text{ for } i < j < k,$$

such that

<u>S 1</u> : Π^{t_i} is a left inverse to Q_i:

$$\Pi^{t_i} Q_i = I_{\mathcal{Y}_i}$$

<u>S 2</u> : Q_i is a left inverse to Π^{t_i}'s restriction to $\mu_i(\Theta)$:

$$Q_i \Pi^{t_i} \mu_i = \mu_i$$

<u>S 3</u> : All diagrams for $i < j$

$$\begin{array}{ccc} \mathcal{X}_i & \xleftarrow{P_{ij}} & \mathcal{X}_j \\ Q_i \uparrow & & \uparrow Q_j \\ \mathcal{Y}_i & \xleftarrow{Q_{ij}} & \mathcal{Y}_j \end{array}$$

are commutative, i.e.

$$Q_i Q_{ij} = P_{ij} Q_j .$$

Note that it then follows that

$$Q_{ij} = \Pi^{t_i} Q_i Q_{ij} = \Pi^{t_i} P_{ij} Q_j .$$

A sufficient system $(t_i)_{i \in I}$ transforms the projective statistical field F into the field G, where

$$G = \{ (Y_i)_{i \in I}, (Q_{ij})_{i<j}, \Theta, (\nu_i)_{i \in I} \}$$

and

$$\nu_i = \Pi^{t_i} \mu_i, \text{i.e.}$$

$$\nu_i (B_i) = \mu_i (t_i^{-1} (B_i)).$$

Note also that the existence of (Q_{ij}) such that $Q_{ij} Q_{jk} = Q_{ik}$ and S 3 holds is equivalent to assuming

$$Q_i \Pi^{t_i} P_{ij} Q_j = P_{ij} Q_j .$$

Note that if $t = (t_i)_{i \in I}$ is sufficient in the field F and $s = (s_i)_{i \in I}$ is sufficient in the field G, the composition

$$(s \circ t)_i = s_i \circ t_i$$

is sufficient in F.

We shall now define the important notion of a canonical statistical field.

<u>2.1 Definition</u> The projective statistical field F is said to be <u>canonical</u> with <u>canonical statistics</u> $(t_i)_{i \in I}$ if

i) $t = (t_i)_{i \in I}$ is sufficient in F

ii) $$\Theta = \varprojlim_{i \in I} \mathcal{Y}_i$$

with canonical projections $\nu_i : \Theta \to \mathcal{Y}_i$.

In other words $\{\nu_i(\theta,\cdot), \theta \in \Theta\}$ shall equal the extreme points of the set of consistent families in the system

$$(\mathcal{Y}_i)_{i \in I},\ (Q_{ij})\ i < j$$

and

$$\theta \to (\nu_i(\theta,\cdot))_{i \in I}$$

shall be a homeomorphism of Θ onto $\mathcal{E}$.

We thus see that the notion of a canonical projective statistical field is a natural generalisation of an extremal family, considered earlier, just that here also a parametrisation of the family is involved.

As an example of such a projective statistical field which is not quite standard, consider the following: Let $I = \mathbb{N}$ and $\mathcal{X}_i = \mathcal{X} \times \ldots \times \mathcal{X}$ (i times) with, say $\mathcal{X} = \{0,1\}$. Think of $\mathcal{X}_i$ as a sample of size i from a population of items, and let 1 on a particular coordinate denote that the item is defective. If $i<j$, $\mathcal{X}_i$ is a <u>subsample</u> of the samples of sizes j. The Markov kernels P_{ij} now characterize the <u>sampling procedure</u> and, for example, in simple random sampling P_{ij} would for every x_j in $\mathcal{X}_j$ give all points in x_i in $\mathcal{X}_i$ such that x_i could be obtained by subsampling from x_j, the same probability.

Letting now $\mathcal{Y}_i = \{0,\ldots,i\}$ and $t_i(x_i)$ equal to the number of defective items in sample i, one can directly verify that S1 and S2 are satisfied. A projective statistical field is given by specifying distributions $m_i(\theta)$, $i \in I$, $\theta \in \Theta$ that are consistent. That the system (t_i) $i \in I$ is sufficient and transitive, can be expressed as the distributions $\nu_i(\theta)$ of the number of defective items are reproducible under simple random sampling, cf. Hald (1960). Hald showed, loc. cit., that the only distributions, reproducible under simple random sampling are mixed binomial distributions. So if we consider the canonical projective statistical field associated with this sampling structure, this is given by assuming $\Theta=[0,1]$ and $\mu_i(\theta)$ letting items be independent and identically distributed with proportion θ of the items being defective. Considerations of this type for more complicated sampling structures are relevant for the model based approach to survey sampling. See also Kingman (1978) for related problems.

3. Canonical projective statistical fields on repetitive structures

We shall here finally take the opportunity to give a precise definition of the notion of a repetitive structure.

A repetitive structure is a projective system of Polish spaces and continuous maps, i.e. consisting of a family

$$(X_i)_{i \in I} \text{ of Polish spaces}$$

and a family $(p_{ij})_{i<j}$ of coherent continuous maps:

$$p_{ij} p_{jk} = p_{ik} \quad i<j<k.$$

The projective limit

$$X = \varprojlim_{i \in I} X_i$$

is the set of consistent sequences

$$\{(x_i)_{i \in I} : p_{ij}(x_j) = x_i\}$$

together with the canonical projections

$$p_i : (x_j)_{j \in I} \to x_i$$

satisfying

$$p_{ij} p_j = p_i \quad \text{for } i<j .$$

In chapter one we investigated canonical fields on repetitive structures of the trivial type where I only had one element.

In chapter two we considered structures of the type where $I = \mathbb{N}$

$$(\underset{\sim}{X}_n = X_1 \times \ldots \times X_n)_{n \in \mathbb{N}}$$

and the projections were given as coordinate projections

$$p_{mn}(x_1,\ldots,x_n) = (x_1,\ldots,x_m) .$$

In chapter III we considered the special case of the above given as

$$\underset{\sim}{X}_n = X \times \ldots \times X \quad \text{(n times)}$$

and coordinate projections.

Note that the spaces here denoted by X_i corresponds to those earlier denoted $\underset{\sim}{X}_i$. In the present chapter we will not use the fact that X_i occasionally are made up by products of simpler spaces.

Such a repetitive structure can also be considered as projective systems of Markov kernels and Polish spaces, where the Markov kernels $\Pi^{p_{ij}}$ are those induced by the maps p_{ij}.

We can thus speak of a projective statistical field on a repetitive structure, and this is then given by a Markov kernel

$$\mu : \Theta \to X$$

where

$$(\mu_i = \Pi^{p_i}\mu)_{i \in I}$$

is the corresponding consistent family and for each $\theta \in \Theta$

$$\mu(\theta,\cdot) \in \mathbb{P}(X)$$

is the measure defined by its consistent family as in Corollary 1.2.

Suppose now that $t_i : \mathcal{X}_i \to \mathcal{Y}_i$ is a sufficient system for this field as defined in the previous section.

Introduce the random variables on $\mathcal{X}$

$$X_i(x) = x_i$$

and let

$$Y_i(x) = t_i(X_i(x)),$$

be the random variables corresponding to the statistic. In analogy with chapter II we have

<u>3.1 Proposition</u> <u>For all</u> $i < j$ X_i <u>and</u> Y_j are <u>conditionally independent given</u> Y_i.

<u>Proof</u> By sufficiency properties S 1 and S 2, Q_i determines the conditional distribution of X_i given Y_i. But, given Y_j, $\Pi^{p_{ij}}Q_j$ determines the conditional distribution of X_i given Y_j. But since we have

$$Q_i \Pi^{t_i} \Pi^{p_{ij}} Q_j = \Pi^{p_{ij}} Q_j$$

Q_i is also the conditional distribution of X_i given Y_i <u>and</u> Y_j. □

<u>3.2 Corollary</u> <u>For any increasing sequence</u> $i_1 < i_2 < \ldots < i_n < \ldots$, <u>the process</u> $Y_{i_1}, Y_{i_2}, \ldots$ <u>has the Markov property</u>.

<u>Proof</u> For all n, X_{i_n} and $Y_{i_{n+1}}$ are conditionally independent given Y_{i_n}. But $Y_{i_1}, \ldots, Y_{i_{n-1}}$ are functions of X_{i_n}. □

Note also, that a statistic $(t_i)_{i \in I}$ is sufficient in the sense defined here <u>exactly when it is sufficient and the conclusion in Proposition 3.1 holds</u>.

Consider now a system $(t_i)_{i\in I}$ of surjective continuous maps

$$t_i : \mathcal{X}_i \to \mathcal{Y}_i$$

and a system of conditional distributions, i.e. of Markov kernels

$$Q_i : \mathcal{Y}_i \to \mathcal{X}_i$$

satisfying S 1:

$$\Pi^{t_i} Q_i = I_{\mathcal{Y}_i}$$

and the consistency condition

$$Q_i \Pi^{t_i} \Pi^{p_{ij}} Q_j = \Pi^{p_{ij}} Q_j .$$

We can then define

$$Q_{ij} = \Pi^{t_i} \Pi^{p_{ij}} Q_j$$

and $(\mathcal{Y}_i)_{i\in I}, (Q_{ij})_{i<j}$ is then a projective system of Polish spaces and Markov kernels, since

$$\begin{aligned} Q_{ij} Q_{jk} &= \Pi^{t_i} \Pi^{p_{ij}} Q_j \Pi^{t_j} \Pi^{p_{jk}} Q_k \\ &= \Pi^{t_i} \Pi^{p_{ij}} \Pi^{p_{jk}} Q_k = \Pi^{t_i} \Pi^{p_{ik}} Q_k = Q_{ik} . \end{aligned}$$

Let now

$$\Theta = \varprojlim_{i\in I} \mathcal{Y}_i$$

and let $\nu_i : \Theta \to \mathcal{Y}_i$ be the canonical projections: $\nu_i(\theta) = \theta_i$.

The family of Markov kernels

$$\mu_i : \Theta \to \mathcal{X}_i$$

given by

$$\mu_i = Q_i \nu_i$$

make now

$$F = \{ (X_i)_{i \in I}, (\Pi^{p_{ij}})_{i < j}, \Theta, (\mu_i)_{i \in I} \}$$

a canonical projective statistical field with $(t_i)_{i \in I}$ as canonical statistic. The <u>extremal</u> family induced by $(t_i)_{i \in I}$ and $(Q_i)_{i \in I}$ is then the family

$$\{\mu(\theta, \cdot), \ \theta \in \Theta\}$$

of probabilities on X given by their consistent families of marginals.

Similarly the <u>maximal</u> family is the family with marginal distributions

$$(Q_i \psi_i), \ \psi \in M_y,$$

where as usual M_y is the space of consistent families of probability measures in the Y- system.

4. Limit theorems for maximal and extremal families on repetitive structures

We shall here generalise the results of chapter II.4 to general repetitive structures.

Consider for any measure μ in the maximal family on a repetitive structure the system of random measures

$$\{M_{ij},\ j > i\}$$

where M_{ij} is the probability measure on X_i given as

$$\begin{aligned} M_{ij}(B_i) &= Q_i Q_{ij}(Y_j, B_i) \\ &= \Pi^{P_{ij}} Q_j(Y_j, B_i) \end{aligned}$$

or, equivalently

$$\int f_i dM_{ij} = E_\mu(f_i(X_i) \mid Y_j) \quad f_i \in C(X_i)$$

and

$$E_\mu \int f_i dM_{ij} = \int f_i d\mu_i .$$

Evidently, for all μ in the maximal family, $(M_{ij})_{j > i}$ are bounded, reverse martingales of measures in the sense that for any $f_i \in C(X_i)$

$$(\int f_i dM_{ij})_{j \geqq i}$$

is a bounded, reverse martingale w.r.t. the decreasing σ-algebras

$$F^j = \sigma\{Y_k,\ k > j\}.$$

Thus, for any f_i, this converges in mean along the directed set and almost surely along any cofinal sequence.

If we now as in chapter II introduce the extended tail σ - algebra A_μ as the σ - algebra generated by subsets A of X satisfying

$$X_i \perp_\mu^{Y_i} 1_A \text{ for all } i \in I$$

and for $\underset{\sim}{i} = (i_n)_{n\in\mathbb{N}}$ being a fixed cofinal sequence the corresponding tail σ - algebra

$$A_{\underset{\sim}{i}} = \bigcap_n \sigma\{Y_{i_N}, N \geqq n\},$$

we get as in chapter II that

$$A_{\underset{\sim}{i}} \subseteqq A_\mu$$

by the Markov property.

We are then ready to show the following limit theorem, analogous to Theorem II 4.4

4.1 Theorem For any μ in the maximal family, the following conditions are equivalent

(i) μ is in the extremal family

(ii) A_μ is μ - trivial

(iii) for any cofinal sequence $\underset{\sim}{i}$

$$M_{ii_n} \xrightarrow[n\to\infty]{w} \mu_i \quad \mu\text{ - almost surely}$$

(iv) there exists a cofinal sequence $\underset{\sim}{i}$ such that

$$M_{ii_n} \xrightarrow[n\to\infty]{w} \mu_i \quad \mu\text{ - almost surely.}$$

Proof Most of the proof is a repetition of the proof of Theorem

II 4.4 and we shall thus only give details where these are essentially different.

(i) ⇒ (ii) If $A \in \mathcal{A}$, $\mu(A) \in]0,1[$, the conditionals $\mu(\cdot|A)$ and $\mu(\cdot|A^c)$ are in the maximal family and

$$\mu(\cdot) = \mu(A) \cdot \mu(\cdot|A) + \mu(A^c) \cdot \mu(\cdot|A^c).$$

(ii) ⇒ (iii) Choose a countable class $\mathcal{D}_i \subseteq C(X_i)$ defining the weak topology on $\mathbb{P}(X_i)$. Then for all $f_i \in \mathcal{D}_i$

$$\int f_i dM_{ii_n} \to M_\infty(f_i) \quad \text{a.s. } \mu$$

defines a $\mathcal{A}_\mu$ - measurable random variable which then has a degenerate distribution and therefore is equal to its expectation and thus

$$\mu_i(f_i) = \lim_{n\to\infty} \int f_i dM_{ii_n} \quad \text{a.s. } \mu$$

Since $\mathcal{D}_i$ is countable, this holds simultaneously for all $f_i \in \mathcal{D}_i$ and we thus have

$$M_{ii_n} \overset{w}{\to} \mu_i \quad \text{a.s. } \mu.$$

(iii) ⇒ (iv) is trivial.

(iv) ⇒ (i) is just a repetition of the argument used to prove (iii) ⇒ (i) in Theorem 4.4. □

Note as a consequence we could as well have looked at

$$N_{ij}(B_i) = Q_{ij}(Y_j, B_i) = \Pi^{t_i} Q_i Q_{ij}(Y_j, B_i) = (\Pi^{t_i} M_{ij})(B_i)$$

and then formulated the statements as

$$N_{ij} \overset{w}{\to} \nu_i \quad \text{a.s. } \nu$$

along any cofinal sequence.

In fact, the almost sure convergence takes place simultaneously for all i, since, if $(i_n)_{n\in\mathbb{N}}$ is cofinal

$$M_{ij} \xrightarrow[j\to\infty]{w} \mu_i \ \forall i \Leftrightarrow M_{i_n j} \xrightarrow[j\to\infty]{w} \mu_{i_n} \ \forall n\in\mathbb{N} .$$

There seems no reason to assume a.s. convergence along the directed set itself in general.

In any examples, we are also quite happy with just taking out a convenient cofinal sequence and showing convergence along such a subsequence.

We define the set $\mathcal{B}_y$ of Boltzmann laws as those elements of $\mathcal{M}_y$ that can be obtained as limits, i.e.

ν is an element of $\mathcal{B}_y$ if there is an $x\in X$ and a cofinal sequence $(i_n)_{n\in\mathbb{N}}$ such that for all $i\in I$

$$\nu_i(\cdot) = \lim_{n\to\infty} Q_{ii_n}(t_{i_n}(x_{i_n}),\cdot) \tag{4.1}$$

where the limit is in the weak topology on $\mathbb{P}(\mathcal{Y}_i)$.

We then have the following important corollary to Theorem 4.1:

4.2 Corollary $\mathcal{B}_y \supseteq \mathcal{E}_y$.

Proof Theorem 4.1 says that the set of x's satisfying (4.1) even have μ-probability one for any extreme point ν. □

As usual we will use this corollary in examples to find $\mathcal{E}_y$ by first finding $\mathcal{B}_y$ and then picking out those that are extreme by investigating their tail behaviour.

In the continuous case it is frequently easier to find B_y than in the discrete case.

Also here we shall in general be interested in investigating problems of the type

$$B = E \ ?$$

5. Poisson Models

An interesting class of examples are associated with the Poisson distribution. Let us first look at the following, taken from Martin-Löf (1974):

5.1 Example (Occupancy numbers)

Consider the case with I=$\mathbb{N}$ and the spaces

$$X_i = \dot{\cup}_{n\in\mathbb{N}}\{1,\ldots,i\}^n$$

An element X_i is to be thought of as an allocation of a number of balls n into a number of cells i, such that each point in X_i contains a number n and a word with n letters from the alphabet $\{1,\ldots,i\}$, specifying for each ball, the cell in which it is placed.

If i<j, we shall think of X_i as being obtained by only observing what is in the first i cells of the cells $\{1,\ldots,j\}$, such that the projection map p_{ij} deletes all balls in cells $\{i+1,\ldots,j\}$. For example, if i=2 and j=4

$$p_{24}(8,11242311)=(6,112211).$$

Suppose we assume that it is a sensible summary of the data to just quote the total number n of balls, i.e. we specify

$$t_i(n_i,a_1\ \ldots\ a_{n_i})=n_i$$

to be sufficient and the conditional distribution to be uniform on X_i (i.e. the microcanonical distribution):

$$q_i(x_i \mid t_i(x_i)=n_i)=i^{-n_i}.$$

Note that this is a genuine projective system in that the projections p_{ij} above are not coordinate projections in a Cartesian product. The extremal model associated with this repetitive structure and the given sufficient statistics has the occupancy numbers $(N_i)_{i\in I}$ to be Poisson distributed with an unknown parameter $\lambda\geq 0$, and the N_i balls then dropped at random into the i cells.

5.2 Example (The multiplicative Poisson model)

An important model for analysing data from e.g. traffic accidents is the multiplicative Poisson model. Here counts of the number of accidents - claims to an automobile insurance company - are partitioned according to, say, two criteria, for example the type of car involved and the month in which it has occurred. Such data can be thought of as occupancy numbers like those in the previous example, if we think of the accidents as supplemented with a list of characteristics of the accidents, identifying e.g. the location in space of the accidents. To simplify our description we shall directly assume that we deal with possibly Poisson distributed numbers. This will determine our choice of conditional distributions Q given the sufficient statistics.

So let I be the set of all subsets of $\mathbb{N}^2$ of the form $A\times B$ with A and B being finite subsets of $\mathbb{N}$. We think of A as a set of car makes and B a set of periods. We let

$$\mathcal{X}_i = \mathcal{X}_{A\times B} = \mathbb{N}^{A\times B}$$

be the sample spaces and usual coordinate projections make up our repetitive structure.

Now it may be considered feasible to summarize such a set of data, by letting

$$t_{A\times B}(Y_{\alpha\beta},\alpha\in A,\beta\in B)=(R_\alpha,S_\beta,\alpha\in A,\beta\in B),$$

where

$$R_\alpha = \sum_\beta Y_{\alpha\beta} \qquad S_\alpha = \sum_\alpha Y_{\alpha\beta}$$

are the total number of accidents for each car make, as well as the total number of accidents in each month.

We further assume that, apart from this, the data has "occupancy structure" such as indicated above, leading to the specification of the conditional distributions as generalized hypergeometric:

$$Q_{A\times B}(Y_{\alpha\beta}=y_{\alpha\beta}|R_\alpha=r_\alpha,S_\beta=s_\beta|(\alpha,\beta)\in A\times B)=\frac{1}{y_{..}!}\prod_{\alpha\beta}\prod(y_{\alpha\beta}!)^{-1}r_\alpha!r_\beta!,$$

where $y_{..}=\Sigma_\alpha\Sigma_\beta y_{\alpha\beta}=\Sigma_\alpha r_\alpha=\Sigma_\beta s_\beta$.

All extreme points for this problem are not known. It has been shown in Martin-Löf (1970) that the Boltzmann laws obtained when passing to the limit in such a fashion that

$r_\alpha s_\beta/y_{..}$ is bounded and bounded away from zero,

will be multiplicative Poisson laws, i.e. the random variables $Y_{\alpha\beta}$ are independent and Poisson distributed with parameters $\lambda_\alpha\mu_\beta,\lambda_\alpha,\mu_\beta>0$ and unknown. These are certainly also extreme and we believe that these, in fact, are the only non-degenerate extreme laws.

An important and interesting modification of this is the following, also taken from Martin-Löf (1970):

<u>5.3 Example</u> (Wilson's traffic model)

Consider the same setup as in Example 5.2, again with the same type of data and sample spaces, with a few modifications:

1. The two classification criteria are 'residential area' and 'area of work'.

2. The data $y_{\alpha\beta}$ are counts of the number of people living in area α and working in area β.

3. With each such number is associated a number $c_{\alpha\beta}$ that represents the <u>cost of transportation</u> from area α to area β.

The row and column sums, defined similarly to those in the example above, correspond here to the total number of people living in, resp. working in, the areas. However, in this case, as indeed done by Wilson (1967), it seems reasonable to supplement these by the statistic

$$T_{A\times B} = \sum_{\alpha\beta}\sum c_{\alpha\beta} Y_{\alpha\beta}$$

which is <u>the total cost of transportation</u>.

We specify the conditional distributions to be those of i.i.d. Poisson variables. Consider now the measures $P_{\underset{\sim}{\lambda},\underset{\sim}{\mu},\underset{\sim}{\nu}}$ where $X_{\alpha\beta}$ are independent and Poisson distributed with

$$E\,X_{\alpha\beta} = \lambda_\alpha \mu_\beta \nu^{c_{\alpha\beta}}$$

to be extreme points. But here the situation is complicated. Whether they are extreme or not depends critically on the behaviour of the costs $c_{\alpha\beta}$. The situation is parallel to that in Example II.9.20, just more difficult. We do not know what the real conditions are.

6. Exponential Families

Here we shall briefly sketch some recent results dealing with exponential families in $\mathbb{R}^k$, in parallel with the discrete ones in our chapter III.

The basic repetitive structure is given by

$$X_i = \mathbb{R}^k \times \ldots \times \mathbb{R}^k \qquad (i \text{ times})$$

and projections are coordinate projections. The sufficient statistics are sums, i.e.

$$t_i(\underset{\sim}{x}_i) = x_1 + \ldots + x_i$$

and the conditional distribution is supposed to have density $q_{ij}(x|y)$ wrt. Lebesgue measure λ equal to

$$q_{ij}(x|y) = \frac{f_0^*(x_1)\ldots f_0(x_i) f^{*(j-i)}(y - \sum_i x_i)}{f_0^{*j}(y)},$$

where we assume that for some j_0, it holds for all $j > j_0$ that f_0^{*j} is bounded, continuous and strictly positive. Then the extremal family generated by this system, is the exponential family $(P_\theta, \theta \in D)$ where, according to P_θ, X_i are independent, identically distributed with density f_θ, where

$$f_\theta(x) = \frac{e^{\theta \cdot x} f_0(x)}{\phi(\theta)}, \tag{6.1}$$

and

$$D = \{\theta | \phi(\theta) < \infty\}.$$

This, together with other results, is proved by Küchler and Lauritzen (1986) and by Diaconis and Freedman (1987b).

We shall here indicate the main line of the proof given in the first of these two references.

First we verify that f_0^{*j} bounded, continuous and positive implies through dominated convergence that the kernels Q_{ij} are continuous, such that the general theory applies.

Then we know that we just have to identify the extreme points of M. Firstly, we see that all measures μ_i, where $(\mu_i)_{i\in\mathbb{N}}\in M$, are absolutely continuous. But this follows from the expression

$$\mu_i(A) = \int Q_{ij}(A|y)\mu_j(dy),$$

because $\lambda(A)=0 \Rightarrow Q_{ij}(A|y)=0 \Rightarrow \mu_i(A)=0$. Then, let h_j be the density of μ_j wrt. f_0^{*j}. Simple manipulations imply that

$$h_i(x) \overset{a.s.}{=} \int h_{i+k}(x+y)f_0^{*k}(y)dy$$

and a smoothing argument gives that the densities can be chosen such that for all x:

$$h_i(x) = \int h_{i+k}(x+y)f_0^{*k}(y)dy.$$

Arguments parallel to those used in section III.1. gives that extreme points must satisfy

$$h_i(x)h_k(y) = h_{i+k}(x+y) \text{ a.e. } \lambda\otimes\lambda$$

and a smoothing argument (integrating over y in a small interval) gives that, in fact, also this equation holds everywhere. Diaconis and Freedman (1987b) base their proof on a lemma, stating that if $X_1,\ldots,X_k$ are i.i.d. and have conditional distribution Q as we defined then their densities look like in (6.1).

7. Examples from continuous time stochastic processes

We shall consider a few simple examples from stochastic processes in continuous time.

7.1 Example (The Wiener process with unknown linear drift).

Let for $T \in [0,\infty[$ the spaces X_T in our repetitive structure be given as

$$X_T = C([0,T])$$

and let the projections simply be restrictions.

Let the consistent family of Markov kernels be defined as

$$\mu_T : \mathbb{R} \to C([0,T]),$$

where $\mu_T(\theta,\cdot)$ is the Wiener process with drift θ, i.e. the Gaussian process with sample functions in $C([0,T])$ having covariance function

$$\mathrm{Cov}_\theta(X(s),X(t)) = s \wedge t$$

and mean value function

$$E_\theta X(t) = \theta t.$$

The projective limit of the spaces $C([0,T])$ is here the space of continuous functions on $[0,\infty[$ with the topology of compact convergence.

Let $Y_T = \mathbb{R}$ and consider the statistics

$$t_T : C([0,T]) \to \mathbb{R}$$

$$t_T(X(t),\ t \in [0,T]) = X(T)$$

The conditional distribution of $(X(t),\ t \in [0,T])$ given $X(T) = x_T$ is the <u>Brownian bridge</u>, the Gaussian process with sample functions in $C([0,T])$, mean function and covariance function

$$EX(t) = \frac{t}{T} \cdot x_T$$

$$\mathrm{Cov}(X(s),\ X(t)) = \frac{s(T-t)}{T} \text{ if } s \leqq t.$$

Combining this with the Markov property of the Wiener process we get that $(t_T)_{T \in [0,\infty[}$ <u>is a sufficient system</u> in the above defined projective statistical field.

When $T \to \infty$, the covariance function tends to the covariance function of the Wiener process.

If the conditional measures shall converge to anything non-degenerate, we must have

$$\frac{x_T}{T} \to \theta$$

and then the mean value function converges to that of the Wiener process with drift θ.

It follows that the Wiener processes with drift exactly are the Boltzmann laws for this problem. On the other hand

$$\frac{X(T)}{T} \xrightarrow{a.s} \theta$$

in these Boltzmann laws, and they are therefore all extreme, and we have $\mathcal{B} = \mathcal{E}$ and that the projective statistical field here defined is canonical.

It follows that the Wiener processes with linear drift have trivial tail σ-algebras.

An example related to the above is the following

7.2 Example (The Ornstein-Uhlenbeck process with hyperbolic sine/cosine drift).

Consider for $a < b \in \mathbb{R}$ the spaces

$$X_{[a,b]} = C([a,b])$$

Let the projections again be restrictions and note that the projective limit is the space of continuous functions on $\mathbb{R}$ with the topology of compact convergence.

Consider the family of distributions for $\alpha,\beta \in \mathbb{R}$ being the Gaussian processes with continuous sample functions, mean value and covariance functions given as

$$E_{\alpha,\beta}X(t) = \alpha \sinh t + \beta \cosh t$$

$$\mathrm{Cov}_{\alpha,\beta}(X(s),X(t)) = e^{-|s-t|}$$

Consider the statistics

$$t_{[a,b]}(X(t),\ t \in [a,b]) = (X(a),\ X(b)).$$

The conditional distribution of $(X(t),\ t \in [a,b])$ given $t_{[a,b]} = (x_a, x_b)$ can after some calculations be found to be the Gaussian process with sample functions in $[a,b]$ and with mean and covariance function

$$EX(s) = \alpha^*_{a,b} \cosh s + \beta^*_{a,b} \sinh s$$

$$\mathrm{Cov}(X(s),X(t)) = e^{s-t} - \frac{(e^{2a-s-t} - e^{2a+s-t-2b} + e^{s+t-2b} - e^{2a-s+t-b})}{1 - e^{2a-2b}} \text{ if } s \leqq t,$$

where α^*, β^* are determined as

$$\alpha^*_{a,b} = \frac{x_a \sinh b - x_b \sinh a}{\sinh (b-a)}$$

$$\beta^*_{a,b} = \frac{x_b \cosh a - x_a \cosh b}{\sinh (b-a)}$$

Thus $t_{[a,b]}$ is sufficient and, by the Markov property of the Ornstein-Ublenbeck process, also sufficient in the sense of a system of transformations of a projective statistical field.

When $[a,b] \uparrow \mathbb{R}$, i.e. $a \to -\infty$, $b \to \infty$, the conditional covariance function tends to that of the Ornstein-Uhlenbeck process and if the mean value function shall tend to something non-degenerate, we must have

$$\alpha^*_{a,b} \to \alpha \quad \beta^*_{a,b} \to \beta.$$

Thus the Ornstein-Uhlenbeck processes with hyperbolic sine/cosine drifts are the Boltzmann laws for this problem.

But they are also extreme since

$$E_{\alpha,\beta}(\hat{\alpha}^*_{a,b}(X(a), X(b)) = \alpha$$

$$E_{\alpha,\beta}(\hat{\beta}^*_{a,b}(X(a), X(b)) = \beta$$

and the corresponding variances are easily seen to tend to zero which implies that $\mathcal{B} = \mathcal{E}$ and the above defined projective statistical field is canonical.

<u>7.3 Example</u>. (Exponential families of processes with stationary and independent increments). This is an extension of Example 7.1, reported by Küchler and Lauritzen (1986). We consider the spaces $\mathcal{X}_T$ = cadlag([0,T],S) of S-valued functions that are right-continuous and have left limits and the sufficient statistic (as in Example 7.1) to be the value at time T, i.e. as in (7.1). Here S is one of the Abelian semigroups $\mathbb{R}^k$, $\mathbb{Z}^k$, $\mathbb{N}$ or $[0,\infty)$, all with usual addition.

Let P^0 be a fixed probability measure on $\mathcal{X}$ giving X_t stationary and independent increments, such that the marginal distribution of X_t according to P^0 is given by a convolution semigroup $(\mu_t^0)_{t>0}$.

Let λ be the invariant measure on S, i.e. Lebesgue measure or counting measure.

Before choosing a specific version of the conditional probabilities $P^0(A|X_t=y)$, we distinguish between the following cases:

<u>Case 1</u>: $S = \mathbb{R}^k$ and $\mu_t^0(\{0\})\equiv 0$ for $t>0$.

In this case we assume that μ_t^0 admits a density p_t^0 w.r.t. λ and that there is a $t^*\geq 0$ such that for each $t>t^*$, p_t^0 can be chosen to be <u>bounded, strictly positive and continuous on</u> S. Note that this is the case if $p_{t^*}^0$ can be chosen to fulfil these requirements, since $p_t^0 = p_{t-t^*}^0 * p_{t^*}^0$ for $t>t^*$.

<u>Case 2</u>: $S = [0,\infty), \mu_t^0(\{0\})\equiv 0$ for $t>0$.

Here we assume that p_t^0 for all $t>0$ can be chosen to be <u>strictly positive on $S\setminus\{0\}$ and continuous there as well</u>. The value of p_t^0 at 0 does not play any role and could be undefined. We further assume that p_t^0 is <u>bounded outside any open interval containing zero for all $t>0$</u>.

In the following cases we assume that $\mu_t^0(\{0\})>0$ for some $t>0$, or, equivalently that $\mu_t^0(\{0\})>0$ for all $t>0$.

Then (X_t) is a so-called compound Poisson process under P^0 (see e.g. Feller (1971) which is piecewise constant and jumps at random times $\tau_1<\tau_2<\ldots<\tau_n<\ldots$, where $(\tau_{i+1}-\tau_i)$ with $\tau_0\equiv 0$ are independent and identically exponentially distributed with some parameter $a>0$.

<u>Case 3</u>: $S = \mathbb{R}^k$.

Here we assume that the size of the jumps of (X_t), which are independent and identically distributed, have a density f w.r.t. Lebesgue measure λ. We also assume f to be <u>strictly positive on $S\setminus\{0\}$ and continuous there as well</u>. Further we assume that f is <u>bounded outside any open rectangle containing zero</u>. We have

$$\mu_t^0(dx) = \varepsilon_0(dx)e^{-at}+p_t^0(x)\lambda(dx),$$

where here and in the sequel, ε_{x_0} is the probability measure degenerate at x_0, and

$$p_t^0(x) = e^{-at}\sum_{k=1}^{\infty}\frac{(at)^k}{k!}f^{*k}(x), \quad x\neq 0,\ t>0.$$

Obviously, p_t^0 satisfies the same regularity conditions as f.

Note that $P^0\{X_t=0,\tau_1>t\} = \exp(-at)$ and that

$$P^0\{X_t\in dx,\tau_1\leq t\} = p_t^0(x)\lambda(dx).$$

Thus μ_t^0 in this case is not absolutely continuous w.r.t. λ.

<u>Case 4</u>: $S = [0,\infty)$

Here we assume the same as in case 3. The reason for distinguishing between cases 3 and 4 is that the results will be slightly different, see table 7.5.

<u>Case 5</u>: $S = \mathbb{Z}^k$.

In this case we assume that $\mu_t^0(\{x\})>0$ for all $x \in S$ and $t>0$.

<u>Case 6</u>: $S = \mathbb{N}_0$.

The same assumptions as in case 5.

Possibly the regularity assumptions could be weakened but they seem not unnatural and many examples are covered.

The distinction between the cases is essential and reflects the different role of the neutral element 0 of the semigroup.

Next we introduce for $0<s<t$ the functions

$$q_{st}(x,y) = \frac{p_s^0(x)\,p_{t-s}^0(y-x)}{p_t^0(y)} \quad \text{on } S\times\{p_t^0>0\}$$

where $p_u^0(z):=0$ when not defined.

Further let us define the kernels $Q_{st}(A|y)$, $0<s<t$, $A \in \mathcal{B}(S)$, $y \in S$ as follows:

<u>Case 1</u>:
$$Q_{st}(dx|y) = q_{st}(x,y)\lambda(dx), \quad x,y\in S$$

<u>Case 2</u>:
$$Q_{st}(dx|y) = \begin{cases} q_{st}(x,y)\lambda(dx) & x\in S,\, y\in S\setminus\{0\} \\ \varepsilon_0(dx) & x\in S,\, y=0 \end{cases}$$

<u>Cases 3 and 4</u>:
$$Q_{st}(dx|y) = \begin{cases} e^{-as}\varepsilon_0(dx)\,\dfrac{p_{t-s}^0(y)}{p_t^0(y)} + e^{-a(t-s)}\,\varepsilon_y(dx)\,\dfrac{p_s^0(y)}{p_t^0(y)} \\ \qquad + q_{st}(x,y)\lambda(dx), \quad x,y\in S,\ y\neq 0 \\ \varepsilon_0(dx) \qquad\qquad , \quad x\in S, \quad y=0. \end{cases}$$

Cases 5 and 6:

$$Q_{st}(\{x\}|y) = \frac{\mu_s^0(\{x\})\mu_{t-s}^0(\{y-x\})}{\mu_t^0(\{y\})}, \quad x,y\in S.$$

Then a version of $P^0(A|X_t=y)$, $A\in F_t$ is in all cases given by

$$P^0(X_{t_1}\in dx_1,\dots,X_{t_n}\in dx_n|X_t=y) =$$

$$\prod_{k=1}^{n-1} Q_{t_k t_{k+1}}(dx_k|x_{k+1})(Q_{t_n t}(dx_n|y),$$

for all $t_1<t_2<\dots<t_n<t$ and $y\in S$. In the sequel, we shall understand <u>this particular version</u> when writing $P^0(A|X_t=y)$. (F_t denotes the σ-algebra generated by $(X_s, s\leqq t)$).

Let $E(P^0)$ denote the exponential family generated by P^0, which is defined to be the measures P^e giving (X_t) stationary and independent increments with X_t having distributions $\mu_t^\rho, \rho\in D(\mu^0)$, where

$$D(\mu^0) = \{\rho\in\hat{S}_+|-\infty<\psi(\rho)<\infty\},$$

$\hat{S}_+$ is the non-negative semicharacters (exponential functions) on S,

$$\psi(\rho) = \log\int_S \rho(x)\mu_1^0(dx)$$

$$\mu_t^\rho(dx) = \rho(x)\exp\{-t\psi(\rho)\}\mu_t^0(dx).$$

If we then let P_0 denote the measure degenerate at zero, Küchler and Lauritzen (1986) show that

<u>7.4 Theorem</u>. <u>Under the assumptions given above we have:</u>

<u>In cases 1, 4, 5 and 6</u>

$$E = E(P^0),$$

<u>whereas in cases 2 and 3</u>

$$\mathcal{E} = E(P^0) \cup \{P_0\},$$

<u>where</u> P_0 <u>is the degenerate measure</u>

$$P_0\{X_t \equiv 0\} = 1.$$

Note that in cases 4 and 6, the measure P_0 also belongs to $\mathcal{M}$ and $\mathcal{E}$. In these cases $S' = [0,\infty)$ or $\mathbb{N}_0$, the function $1_{\{0\}}(x)$ is in $D(\mu^0)$ and therefore $P_0 \in E(P^0)$.

In cases 1 and 5, P_0 does not belong to $\mathcal{M}$ or $\mathcal{E} = (P^0)$, corresponding to the fact that $1_{\{0\}}(x)$ is not a semicharacter in these semigroups ($S = \mathbb{R}^k$ or $\mathbb{Z}^k$).

In case 2, $1_{\{0\}}(x)$ is a semicharacter, but not in $D(\mu^0)$ because it integrates to zero w.r.t. μ_t^0.

In case 3, P_0 comes into the picture because 0 is the particular state where the process begins at t=0, stays some time, and for each fixed t>0 never hits again with positive probability, once it has left it. So the appearance of P_0 in this case is due to the role of 0 as the <u>initial state</u> rather than the neutral element of the semigroup.

An overview of the results is given overleaf in Table 7.5.

Table 7.5

An Overview of the Assumptions and Results in the Six Different Cases

Case	1	2	3	4	5	6
$\mu_t^0(\{0\})$	0	0	>0	>0	>0	>0
S	$\mathbb{R}^k$	$[0,\infty)$	$\mathbb{R}^k$	$[0,\infty)$	$\mathbb{Z}^k$	$\mathbb{N}_0$
Assumptions	p_t^0 satisfies A1, $t>t^*$	p_t^0 satisfies A2 for all t	f satisfies A2	f satisfies A2	$\mu_t^0(x)>0$ $x\in S$ for all t	$\mu_t^0(x)>0$ $x\in S$ for all t
$P_0 \in M$	no	yes	yes	yes	no	yes
$1_{\{0\}} \in \hat{S}_+$	no	yes	no	yes	no	yes
$1_{\{0\}} \in D(\mu^0)$	no	no	no	yes	no	yes

A1: g is bounded, strictly positive and continuous on S.

A2: g is strictly positive on $S\setminus\{0\}$ and continuous there as well. g is bounded outside any neighbourhood of 0.

We shall not give the proof here but just mention that the line of argument is as that sketched in Section 6.

Note that Example 7.1 belongs to this class of examples. See Küchler and Lauritzen (1986) for many other examples. Here we just give the following

<u>7.6 Example</u>. (<u>A Compound Poisson Process</u>)

Here we let $S = [0,\infty)$, fix $\beta>0$ and $c>0$, and

$$\mu_t^0(dx) = \varepsilon_0(dx)\, e^{-at} + \sum_{k=1}^{\infty} \frac{(at)^k}{k!} \frac{\beta^{kc} x^{kc-1}}{\Gamma(kc)}\, e^{-(a+\beta)x}\, dx,$$

i.e. the process has gamma distributed jumps.

In this example, semicharacters are either usual exponential functions or the function $1_{\{0\}}$ and it is easier for us to work with the set Δ and the function ϕ, where

$$\Delta = \{\delta \in \mathbb{R} \,|\, \phi(\delta) = \log \int e^{\delta x} \mu_1^0(dx) < \infty\},$$

the connection being that if $\rho \neq 1_{\{0\}}$,

$$\rho(x) = e^{\delta x}, \quad \phi(\delta) = \psi(\rho).$$

Our assumptions hold as in case 4 and the minimal normalized space-time harmonic functions for (X_t) are given by

$$\{h_t^\delta(x) = e^{\delta x - \phi(\delta) t} \,|\, \delta \in \Delta\},$$

where $\Delta = (-\infty, \beta)$ and

$$\phi(\delta) = -a\left(1 - \left(1 - \frac{\delta}{\beta}\right)^c\right).$$

The elements P^δ of $\mathcal{E}$ are compound Poisson processes with jump intensities $a(1-\delta/\beta)^c$ and the jumps gamma distributed with parameters c and $\beta-\delta$.

<u>7.7 Example</u>. (<u>The Yule Process</u>)

This example is inspired by Keiding (1974) and the extreme points have recently been found by Jensen and Johansson (1988), where further details can be seen.

We consider $\mathcal{X}_T$, $T>0$ as the space of all non-exploding counting process trajectories. We now first consider the family of Yule processes on $\mathcal{X}_T$, $\{P_\alpha, \alpha>0\}$, where the counting process (N_t) has P_α-intensity equal to $\alpha\cdot(N_{t-}+1)$. As shown by Keiding (1974), the minimal sufficient statistic for (P_α) restricted to $\mathcal{X}_T$, is the pair (N_T, S_T), where

$$S_T = \int_0^T (N_t+1)dt.$$

The conditional distribution of (N_t) given $(N_T, S_T) = (n,y)$ can, see Jensen and Johansson, be described as a counting process.

It was pointed out by Keiding (1974) that this is <u>not</u> an extremal family, which follows from the fact that the random variable $W = \lim N_t P^{-\alpha t}$ has a non-degenerate distribution, implying that the tail σ-algebra of (N_t, S_t) is not trivial.

Consider the following family of probability measures on $\mathcal{X}$. For $\alpha \geqq 0$, $b \in \mathbb{R}$, let $P_{\alpha,\beta}$ be the counting process with intensity equal to $\alpha e^{\beta t}$. If $\beta<0$, then, w.r.t. $P_{\alpha,\beta}$, $N_\infty = \lim_{t\to\infty} N_t$ has a Poisson distribution. Further, if we let $Y_t = (n+1)t - S_t$, then, given $N_\infty = n$, $Y_\infty^n = \lim_{t\to\infty} Y_t^n$ has a $\Gamma(n,-\beta)$ distribution. We can now calculate the distributions

$$Q_{n,a}(\cdot) = P_{1,-1}(\cdot \,|\, N_\infty = n,\ Y_\infty^n = \alpha).$$

It has been conjectured by the author for a long time and shown by Jensen & Johansson (1988) (Theorem 1.3) that

$$\mathcal{E} = \mathcal{B} = \{P_{\alpha,\beta}, \alpha \geqq 0, \beta \geqq 0\} \cup \{Q_{n,a}, n \geqq 1, a>0\}.$$

The proof is complicated and omitted here.

The statistical consequence of the result is to perform inference in one sample from the Yule-process by conditioning on W=w, treating w as a parameter. This possibility was discussed by Keiding (1974) and Basawa and Scott (1984) see also Sweeting (1986).

8. Linear normal models

We shall first recall a well known result about the conditional distribution of $X_1,\ldots X_n$ given $X_1^2+\ldots+X_N^2 = s$, when X_i are i.i.d. $N(0,\sigma^2)$, derived for example by Martin-Löf (1970), see also Cramér (1946) p. 121. The quoted distribution has density

$$f(x_1,\ldots,x_n) = \frac{\Gamma\left(\frac{N}{2}\right)}{\sqrt{\pi}^n\,\Gamma\left(\frac{N-n}{2}\right)}\;\frac{\left(s - \sum_{i=1}^{n} x_i^2\right)^{\frac{N-n}{2}-1}}{s^{\frac{N}{2}-1}}\;1_{[0,s]}\left(\sum_1^n x_i^2\right)$$

Using this result and an appropriate choice of coordinate systems one easily finds the distribution of $Y = AX$ given $BX = b$ and $\|(I_p - B)X\|^2 = s^2$ where

$$X \sim N_p(0,\sigma^2 I_p)$$

A is a $q \times p$ matrix with $AA^* = I_q$,

B is a projection matrix (symmetric and idempotent) with rank r,

$A(I_p - B)$ has full rank $q < p - r$, to be given by the density

$$f(y\,|\,BX = b, \|X - BX\|^2 = s) = \frac{\Gamma\left(\frac{p-r}{2}\right)\left(s - (y - Ab)^*\Sigma^{-1}(y - Ab)\right)^{\frac{p-r-q}{2}-1}}{\sqrt{\pi}^q\,\Gamma\left(\frac{p-r-q}{2}\right)\, s^{\frac{p-r}{2}-1}\,|\Sigma|^{\frac{1}{2}}}$$

on the set where

$$(y - Ab)^*\Sigma^{-1}(y - Ab) \leqq s,$$

* denoting transpose and

$$\Sigma = A(I - B)A^* = I - ABA^*.$$

We shall abstain from going through the calculations.

We shall now proceed to describe a canonical projective statistical field made up by linear normal models:

Let I be a partially ordered set as usual, directed to the right and containing a cofinal sequence.

Let now

$$X_i = \mathbb{R}^{n_i}$$

with $n_i < n_j$ for $i < j$, $n_i < \infty$.

As projections we choose a coherent system of linear maps A_{ij}, satisfying

$$A_{ij} A_{ij}^* = I_i$$

and the usual coherence condition:

$$A_{ij} A_{jk} = A_{ik} \text{ for } i < j < k.$$

The projective limit is $\mathbb{R}^{\mathbb{N}}$ with the usual product topology, and we let A_i denote the canonical projections.

What we aim to construct, is a coherent system of usual linear models, and we let thus

$$L_i \subset X_i$$

be linear subspaces of $X_i = \mathbb{R}^{n_i}$, each having dimension r_i.

We want these to be connected in a consistent way as well, so we assume

$$A_{ij}(L_j) = L_i \quad i < j.$$

Thus the restrictions $\overline{A}_{ij}$ of A_{ij} to L_j together with $(L_i)_{i\in I}$ form again a projective system and we let

$$L = \lim_{\substack{\longleftarrow \\ i\in I}} L_i$$

with corresponding canonical projections $\overline{A}_i$. For $\xi \in L$ we let

$$\xi_i = \overline{A}_i \xi \ .$$

We form now a projective statistical field by taking as parameter space

$$\Theta = L \times [0,\infty[$$

and to $\theta = (\xi,\sigma^2)$ we associate the family of normal distributions on X_i, given as

$$X_i \sim N_{n_i}(\xi_i , \sigma^2 I_i).$$

This family is consistent since

$$X_j \sim N_{n_j}(\xi_j , \sigma^2 I_j) \Rightarrow A_{ij}X_j \sim N_{n_i}(A_{ij}\xi_j , \sigma^2 A_{ij}A_{ij}^*) = N_{n_i}(\xi_i , \sigma^2 I_i)$$

by the assumptions made earlier.

For each $i \in I$ the statistic

$$t_i : X_i \to Y_i = L_i \times [0,\infty[$$

given as

$$t_i(X_i) = (B_i X_i , \|X_i\|^2)$$

where B_i is the orthogonal projection onto L_i, is known to be sufficient, and it is also a sufficient system in the sense here defined, which can be seen by the following argument. We know that

$$A^*_{ij}(\mathcal{X}_i) = A^{-1}_{ij}(0)^{\perp}$$

we can then write X_j as

$$X_j = A^*_{ij}X_i + (X_j - A^*_{ij}X_i) = Y_j + Z_j$$

a sum of independent random variables.

Further, since $L_i = A_{ij}L_j$, $B_iA_{ij}B_j = A_{ij}B_j$ and

$$B_jX_j = B_jA^*_{ij}X_i + B_jZ_j$$

$$= B_jA^*_{ij}(B_iX_i) + B_jZ_j$$

and

$$\|X_j\|^2 = \|Y_j\|^2 + \|Z_j\|^2 = \|X_i\|^2 + \|Z_j\|^2$$

we have

$$T_j = f_{ij}(T_i, Z_j),$$

and since X_i and Z_j are independent, they are conditionally independent given T_i whereby X_i and T_j are independent given T_i.

Suppose now that the following conditions are fulfilled

i) $$n_i - r_i \to \infty \quad \text{for } i \to \infty$$

ii) $$A_{ij}B_jA^*_{ij} \to 0 \text{ for } j \to \infty .$$

The first condition is basically saying that the degrees of freedom available to estimate σ^2 shall tend to infinity as i increases.

The second condition says that the maximum likelihood estimate of ξ_i based on observation of X_j shall be consistent for $j \to \infty$. This is so because

$$B_j X_j$$

is the maximum likelihood estimate of ξ_j based on X_j, such that

$$A_{ij} B_j X_j$$

is the maximum likelihood estimate of ξ_i based on X_j and this has covariance matrix equal to

$$\sigma^2 (A_{ij} B_j B_j^* A_{ij}^*) = \sigma^2 A_{ij} B_j A_{ij}^* \ .$$

It is obvious that we cannot expect the projective statistical field to be canonical <u>unless</u> these conditions are fulfilled. On the other hand we shall show that the conditions are also sufficient. Let for fixed i,j be so large that

$$I_i - A_{ij} B_j A_{ij}^* \quad \text{is regular}$$

which is the case if and only if

$$A_{ij}(I_j - B_j) \quad \text{has full rank.}$$

The conditional distribution of X_i given $t_j(X_j) = (b_j, s_j + \|b_j\|^2)$ is then equal to the distribution with density

$$f_{ij}(x|b_j,s_j) = \frac{\Gamma\left(\frac{n_j - r_j}{2}\right)\left(s_j - (x_i - b_{ij})^* \Sigma_{ij}^{-1}(x_i - b_{ij})\right)^{\frac{n_j - r_j - n_i}{2} - 1}}{\sqrt{\pi}^{n_i}\,\Gamma\left(\frac{n_j - r_j - n_i}{2}\right) s_j^{\frac{n_j - r_j}{2} - 1} |\Sigma_{ij}|^{\frac{1}{2}}}$$

where

$$\Sigma_{ij} = I_i - A_{ij} B_j A_{ij}^*$$

and

$$b_{ij} = A_{ij} b_j .$$

Since $A_{ij} B_j A_{ij}^* \to 0$ and $n_j - r_j \to \infty$, the asymptotic behaviour of this density is the same as

$$f \sim \frac{1}{\sqrt{2\pi}^{n_i}} \left(\frac{n_j - r_j - n_i}{s_j} \right)^{\frac{n_i}{2}} \left(1 - \frac{(x_i - b_{ij})^* \Sigma_{ij}^{-1} (x_i - b_{ij})}{s_j} \right)^{\frac{n_j - r_j}{2} - 1}$$

The third factor in the product is always bounded by one. Thus, if $s_j/(n_j - r_j) \to \infty$, f converges uniformly to zero and we do not get a probability distribution.

The interesting cases occur if $s_j/(n_j - r_j) \to \sigma^2$ where $\sigma^2 \in [0,\infty[$. Let us first consider the case $\sigma^2 = 0$. If $b_{ij} \to b_i$, f tends to infinity at b_i and to zero uniformly outside spheres containing b_i implying that we as limit get the one point measure in b_i. If b_{ij} tends to infinity in the sense that it disappears from any compact subset, the mass disappears and we do not get a limit point. If $\sigma^2 \neq 0$, the mass again disappears unless $b_{ij} \to b_i$ in which case the density converges uniformly on compact sets to

$$\frac{1}{\sqrt{2\pi\sigma^2}^{n_i}} e^{-\frac{\|x_i - b_i\|^2}{2\sigma^2}}$$

i.e. a normal density already in our model.

Therefore the Boltzmann laws are exactly those considered in our projective statistical field. But, as earlier mentioned, in these models

$$A_{ij} X_j \xrightarrow{\text{a.s.}} \xi_i$$

and since $\|X_j - B_j X_j\|^2$ is $\sigma^2\chi^2$ - distributed with $n_j - r_j$ degrees of freedom we also have

$$\frac{1}{n_j - r_j}\|X_j - B_j X_j\|^2 \xrightarrow{a.s.} \sigma^2$$

such that the Boltzmann laws are indeed all extreme. Thus, in this case we also have $\mathcal{B} = \mathcal{E}$ and the projective statistical field is canonical.

As a special case of this, consider the following example:

8.1 Example (Two way analysis of variance)

Let α,β be finite subsets of $\mathbb{N}$ and consider the projective system

$$X_{\alpha,\beta} = \mathbb{R}^{\alpha\times\beta}$$

where

$$(\alpha,\beta) < (\alpha^*,\beta^*) \Leftrightarrow \alpha \subseteqq \alpha^*, \beta \subseteqq \beta^*.$$

Let

$$L_{\alpha,\beta} \cong \mathbb{R}^{\alpha} + \mathbb{R}^{\beta}$$

$$= \{(x_{ij}) : x_{ij} = \xi_i + \beta_j\}$$

Be the usual space of additivity and let the projections be coordinate projections.

We have as is well known

$$(B_{\alpha,\beta}x)_{ij} = \frac{1}{|\beta|}\sum_{j\in\beta} x_{ij} + \frac{1}{|\alpha|}\sum_{i\in\alpha} x_{ij} - \frac{1}{|\alpha||\beta|}\sum_{i\in\alpha}\sum_{j\in\beta} x_{ij}$$

$$= \bar{x}^{\beta}_{i\cdot} + \bar{x}^{\alpha}_{\cdot j} - \bar{x}^{\alpha\beta}_{\cdot\cdot}\ .$$

Also

$$\dim X_{\alpha,\beta} = n_{\alpha,\beta} = |\alpha||\beta|$$

$$\dim L_{\alpha,\beta} = r_{\alpha,\beta} = |\alpha| + |\beta| - 1$$

such that

$$n_{\alpha,\beta} - r_{\alpha,\beta} = (|\alpha| - 1)(|\beta|-1)$$

diverges to infinity if and only if <u>either</u> $|\alpha| \to \infty$ <u>or</u> $|\beta| \to \infty$, provided the other is greater than one.

The 'critical' matrix acts on (x_{ij}) as

$$A_{(\alpha,\beta),(\alpha^*,\beta^*)}\ B_{\alpha^*,\beta^*}\ A^*_{(\alpha,\beta),(\alpha^*,\beta^*)}(x_{ij})$$

$$= \frac{1}{|\beta^*|}\sum_{j\in\beta} x_{ij} + \frac{1}{|\alpha^*|}\sum_{j\in\alpha} x_{ij} - \frac{1}{|\alpha^*||\beta^*|}\sum_{i\in\alpha}\sum_{j\in\beta} x_{ij}$$

$$= \frac{|\beta|}{|\beta^*|}\cdot \bar{x}^{\beta}_{i.} + \frac{|\alpha|}{|\alpha^*|}\bar{x}^{\alpha}_{.j} - \frac{|\alpha||\beta|}{|\alpha^*||\beta^*|}\bar{x}^{\alpha,\beta}_{..}$$

and this converges to zero if and only if <u>both</u> $|\alpha^*| \to \infty$ and $|\beta^*| \to \infty$.

Thus, a projective system of two way classifications as above is canonical if and only if <u>both</u> the number of rows <u>and</u> the number of columns increase to infinity.

Consider now the case where α is fixed throughout, i.e. when only the number of columns tend to infinity. We obtain then the canonical model by <u>conditioning</u> on all the column sums

$$\bar{x}^{\alpha}_{\cdot j} = \frac{1}{|\alpha|}\sum_{i\in\alpha} x_{ij}$$

corresponding to the situation in the conditional Rasch model considered in example II 9.7. To see this, we have to choose new orthonormal systems such that the distribution of $(\bar{X}_{ij})_{i\in\alpha}$ given $\bar{X}^{\alpha}_{\cdot j}$

can be represented by a regular normal distribution.

Let us finally consider

8.2 Example (Linear regression analysis).

For $n \in \mathbb{N}$, let $(t_1,\ldots,t_n,\ldots)$ be a sequence of known real numbers and let our projective system be

$$X_n = \mathbb{R}^n$$

connected with usual coordinate projections.

Consider the consistent family of linear models given as

$$(X_1,\ldots,X_n) \text{ independent, normal with } X_i \sim N(\alpha + \beta t_i, \sigma^2)$$

$$(\alpha,\beta) \in L = \mathbb{R}^2 \text{ unknown an } \sigma^2 \geqq 0 \text{ unknown.}$$

The degrees of freedom are here tending to infinity and the projections are well known to be given as

$$(B_n x)_i = \bar{x}_n + \frac{SPD_{xt}}{SSD_t}(t_i - \bar{t}_n).$$

The critical matrix $A_{nN}B_N A^*_{nN}$ acts on a vector $(x_1,\ldots,x_n)$ as

$$(A_{nN}B_N A^*_{nN}x)_i = \frac{n}{N}\bar{x}_n + \frac{SP^n_{xt} - n\bar{x}_n\bar{t}_N}{SSD^N_t}(t_i - \bar{t}_N)$$

and this tends to zero if and only if

$$N \to \infty,\ SSD^N_t \to \infty,\ \frac{\bar{t}_N}{SSD^N_t} \to 0,\ \frac{\bar{t}^2_N}{SSD^N_t} \to 0\,.$$

The first condition is automatically satisfied, and we intend to show that the two latter are consequences of the condition

$$\mathrm{SSD}_t^N \to \infty .$$

If $|\bar{t}_N|$ is bounded, this is obvious. Note that

$$\frac{\mathrm{SSD}_t^N}{\bar{t}_N^2} = \sum_{i=1}^{N} \left(\frac{t_i}{\bar{t}_N} - 1 \right)^2 .$$

If $|\bar{t}_N|$ tends to infinity, each of the terms in this sum tend to 1 and since the number of terms is increasing, the entire sum tends to $+\infty$ implying

$$\frac{\bar{t}_N^2}{\mathrm{SSD}_t^N} \to 0 .$$

Finally
$$\frac{|\bar{t}_N|}{\mathrm{SSD}_t^N} = \sqrt{\frac{1}{\mathrm{SSD}_t^N} \cdot \frac{\bar{t}_N^2}{\mathrm{SSD}_t^N}}$$

whereby also $\bar{t}_N/\mathrm{SSD}_t^N \to 0$.

Thus, the necessary and sufficient condition for the above projective statistical field to be canonical is

$$\mathrm{SSD}_t^N \underset{N\to\infty}{\to} \infty .$$

9. The Rasch model for item analysis

As our last example we shall consider the Rasch model in the case where both the number of items and the number of persons are allowed to tend to infinity.

This model is of course a discrete version of the two-way analysis of variance but far more complicated. This is partly due to the fact that the discreteness prevents any explicit expressions for the conditional distributions, partly to the fact that the variances of the discrete variables in a non trivial way depend on the expectations.

We have earlier considered the Rasch model for a fixed number of items and even in that situation it was a rather complicated task to describe all the extreme points.

In the present situation we are unfortunately not able to give a complete solution of the problem of finding all the extreme points. Instead we shall give a necessary and a sufficient condition for the usual probability measures in the Rasch model to be extreme.

Consider the doubly infinite array $(X_{ij})_{(i,j)\in \mathbb{N}\times\mathbb{N}}$ of random variables taking values in $\{0,1\}$ and let $P_{\underset{\sim}{\alpha},\underset{\sim}{\beta}}$ $\underset{\sim}{\alpha} = (\alpha_1,\dots,\alpha_n,\dots)$ $\underset{\sim}{\beta} = (\beta_1,\dots,\beta_n\dots)$ be the probability measure on the infinite product space letting X_{ij} be independent and

$$P_{\underset{\sim}{\alpha},\underset{\sim}{\beta}}\{X_{ij}=1\} = \frac{\alpha_i\beta_j}{1+\alpha_i\beta_j} = 1 - P_{\underset{\sim}{\alpha},\underset{\sim}{\beta}}\{X_{ij}=0\}.$$

Consider the projective system of spaces indexed by pairs (A,B) of finite subsets of $\mathbb{N}$, partially ordered as $(A,B) < (A^*,B^*) \Leftrightarrow A \subseteq A^*$, $B \subseteq B^*$ and let

$$X_{A,B} = \{0,1\}^{A \times B}$$

connected with coordinate projections and let

$$P^{A,B}_{\underset{\sim}{\alpha},\underset{\sim}{\beta}}$$

denote the consistent family of marginal distributions. The sufficient system of statistics when $\underset{\sim}{\alpha}, \underset{\sim}{\beta}$ are unknown is given as

$$t_{A,B}((x_{ij})) = (R^B_i,\ i \in A,\ S^A_j,\ j \in B)$$

where R and S are row and column sums

$$R^B_i = \sum_{j \in B} x_{ij},\quad S^A_j = \sum_{i \in A} x_{ij}.$$

The corresponding conditional distributions are the uniform distributions on the set of 0 - 1 matrices with given row and column sums. Assume now for simplicity that $\alpha_i > 0, \beta_j > 0\ \forall i,j$, and consider the two conditions below

$$C_1:\quad \sum_{i=1}^{\infty} \frac{\alpha_i}{(1+\alpha_i)^2} = \sum_{j=1}^{\infty} \frac{\beta_j}{(1+\beta_j)^2} = \infty$$

$$C_2:\quad \sum_{i=1}^{\infty} \frac{\alpha_i \beta_i}{(1+\alpha_i)(1+\beta_i)(1+\alpha_i\beta_i)} = \infty$$

And note first the following fact:

<u>9.1 Lemma</u> $C_2 \Rightarrow C_1$.

<u>Proof</u> We have

$$\frac{\alpha}{(1+\alpha)^2} - \frac{\alpha\beta}{(1+\alpha)(1+\beta)(1+\alpha\beta)} = \frac{\alpha(1+\beta^2)}{(1+\alpha)^2(1+\beta)(1+\alpha\beta)} > 0. \qquad \square$$

Note also that C_1 <u>does not imply</u> C_2, take for example $\alpha_i = \beta_i = i$, then

$$\Sigma \frac{i}{(1+i)^2} = \infty$$

but

$$\Sigma \frac{i^2}{(1+i)^2(1+i^2)} = \Sigma \frac{i}{(1+i)^2} \frac{i}{(1+i^2)} < \infty .$$

We have

9.2 Proposition <u>Condition</u> C_1 <u>is necessary for</u> $P_{\underset{\sim}{\alpha},\underset{\sim}{\beta}}$ <u>to be an extreme point</u>.

Proof Suppose $\Sigma \frac{\alpha_i}{(1+\alpha_i)^2} < \infty$. Then

$$Y_{j\infty} = \sum_{i=1}^{\infty} (X_{ij} - \frac{\alpha_i\beta_j}{1+\alpha_i\beta_j})$$

is a non-degenerate,'tail-measurable'random variable with variance

$$0 < V(Y_{j\infty}) = \sum_{i=1}^{\infty} \frac{\alpha_i\beta_j}{(1+\alpha_i\beta_j)^2} < \infty . \qquad \square$$

We have further:

9.3 Proposition <u>Condition</u> C_2 <u>is sufficient for</u> $P_{\underset{\sim}{\alpha},\underset{\sim}{\beta}}$ <u>to be an extreme point</u>.

<u>Proof</u> By Theorem 4.1 we just have to choose a cofinal sequence and show that the corresponding Markov chain of sufficient statistics has trivial tail σ-algebra.

We shall as usual do this by a coupling argument. Choose the cofinal sequence

$$(A_n, B_n) = (\{1,\dots,n\},\{1,\dots,n\})$$

of quadratic submatrices. Let n be fixed and let

$$(r_1,\dots,r_n;\ s_1,\dots,s_n),(r_1',\dots,r_n';\ s_1',\dots,s_n')$$

be two different vectors of row-and column sums.

Since couplability is an equivalence relation and we have symmetry between rows and columns and the labelings we need only consider the following cases:

i) $s_j' = s_j\ \forall j\ ;\ r_1' = r_1 + 1,\ r_2' = r_2 - 1,\ r_i' = r_i$ for $i = 3,\dots,n$

ii) $s_1' = s_1 + 1,\ s_j' = s_j,\ j = 2,\dots,n$ and

$r_1' = r_1 + 1,\ r_i' = r_i$ for $i = 2,\dots,n$.

In case i) we do the coupling as follows:

Let for $N > n\,(X_{1N}, X_{2N})_{N \geqq n}$ and $(\tilde{X}_{1N}, \tilde{X}_{2N})_{N \geqq n}$ be two systems of independent random variables with

$$P\{X_{1N} = x_{1N},\ X_{2N} = x_{2N}\} = \frac{(\alpha_1 \beta_N)^{X_{1N}} (\alpha_2 \beta_N)^{X_{2N}}}{(1 + \alpha_1 \beta_N)(1 + \alpha_2 \beta_N)},$$

and

$$P\{\tilde{X}_{1N} = 1, \tilde{X}_{2N} = 1 \mid X_{1N} = 1, X_{2N} = 1\} = P\{\tilde{X}_{1N} = 0, \tilde{X}_{2N} = 0 \mid X_{1N} = 0, X_{2N} = 0\} = 1$$

and

$$P\{\tilde{X}_{1N} = \tilde{x}_{1N},\ \tilde{X}_{2N} = \tilde{x}_{2N} \mid X_{1N} + X_{2N} = 1\} = \frac{\alpha_1^{\tilde{x}_{1N}} \alpha_2^{\tilde{x}_{2N}}}{\alpha_1 + \alpha_2} \quad \text{if } \tilde{x}_{1N} + \tilde{x}_{2N} = 1$$

and zero otherwise.

Clearly $(\tilde{X}_{1N}, \tilde{X}_{2N})$ and (X_{1N}, X_{2N}) are identically distributed. Define now further (X_{ij}) and $(\tilde{X}_{ij})$ for $i = 3,\ldots,n;\ j = n+1,\ldots$ and $i = n+1,\ldots\ \ j = 1,2,\ldots$ by letting

$$\tilde{X}_{ij} = X_{ij}$$

and X_{ij} be independent and independent of the foregoing with

$$P\{X_{ij} = 1\} = 1 - P\{X_{ij} = 0\} = \frac{\alpha_i \beta_j}{1 + \alpha_i \beta_j} .$$

Let now T_N and $\tilde{T}_N$ be the set of row and column sums

$$T_N = (R_{1N},\ldots,R_{NN}; S_{1N},\ldots,S_{NN})$$

$$\tilde{T}_N = (\tilde{R}_{1N},\ldots,\tilde{R}_{NN}; \tilde{S}_{1N},\ldots,\tilde{S}_{NN})$$

defined recursively as

$$T_n = (r_1,\ldots,r_n;\ s_1,\ldots,s_n)$$

$$\begin{aligned}\tilde{T}_n &= (r_1',\ldots,r_n';\ s_1',\ldots,s_n') \\ &= (r_1 + 1, r_2 - 1, r_3,\ldots,r_n;\ s_1,\ldots,s_n)\end{aligned}$$

and

$$\begin{aligned}T_{N+1} = (R_{1N} + X_{1,N+1},\ldots,R_{NN} + X_{N,N+1}, \sum_{j=1}^{N+1} X_{N+1,j}; \\ S_{1N} + X_{N+1,1},\ldots,S_{NN} + X_{N+1,N}, \sum_{i=1}^{N+1} X_{i,N+1})\end{aligned}$$

and $\tilde{T}_{N+1}$ correspondingly with ~ over all the symbols.

Then T_N and $\tilde{T}_N$ are both Markov chains with the right transition probabilities.

Further, for all N, $i = 3,\ldots,N$, $j = 1,\ldots,n$

$$R_{iN} = \tilde{R}_{iN},\ S_{iN} = \tilde{S}_{iN}$$

whereas

$$R_{1N} - \tilde{R}_{1N} = \tilde{R}_{2N} - R_{2N} = D_N,$$

where, D_N, watched only at jump times is a symmetric simple random walk. Letting now

$$\tau = \inf\ \{N : D_N = 0\},$$

τ is finite with probability one if and only if D_N jumps infinitely often with probability one , since such a random walk is recurrent.

But the jump probabilities are

$$P\{D_N \neq D_{N-1}\} = P\{X_{1N} \neq \tilde{X}_{1N}\}$$

$$= 2P\{X_{1N} = 1,\ \tilde{X}_{1N} = 0\,|\,X_{1N}+X_{2N}=1\}\cdot P\{X_{1N}+X_{2N}=1\}$$

$$= 2\ \frac{\alpha_1}{\alpha_1 + \alpha_2}\ \frac{\alpha_2}{\alpha_1 + \alpha_2}\ \frac{(\alpha_1 + \alpha_2)\beta_N}{(1 + \alpha_1\beta_N)(1 + \alpha_2\beta_N)}$$

$$= \frac{2\alpha_1\alpha_2\beta_N}{(1 + \alpha_1\beta_N)(1 + \alpha_2\beta_N)(\alpha_1 + \alpha_2)}\ .$$

But, since condition C_2 implies C_1, we have

$$\sum_N \frac{\beta_N}{(1 + \beta_N)^2} = \infty$$

and by the inequality

$$\frac{1}{1+\alpha\beta} \geqq \begin{cases} \dfrac{1}{\alpha(1+\beta)} & \text{if } \alpha \geqq 1 \\[2ex] \dfrac{1}{1+\beta} & \text{if } \alpha < 1 \end{cases} \tag{9.1}$$

we have

$$\sum_N P\{\tilde{X}_{1N} \neq X_{1N}\} = \infty .$$

The Borel-Cantelli lemma now ensures that we get an infinite number of jumps and we therefore have

$$P\{\tau < \infty\} = 1.$$

Defining now

$$T_N' = \begin{cases} \tilde{T}_N & \text{for } N \leqq \tau \\ T_N & \text{for } N > \tau \end{cases}$$

we get that $(T_N, T_N')_{N \geqq n}$ is a successful coupling.

Note that the above coupling could be made with just the weaker assumption C_1.

In case two we have to consider triplets

$$(X_{1N}, X_{N1}, X_{NN}) \text{ and } (\tilde{X}_{1N}, \tilde{X}_{N1}, \tilde{X}_{NN})$$

where (X_{1N}, X_{N1}, X_{NN}) are all independent with the right probabilities and

$$(\tilde{X}_{1N}, \tilde{X}_{N1}, \tilde{X}_{NN}) = (X_{1N}, X_{N1}, X_{NN})$$

<u>unless</u>

$$(X_{1N},X_{N1},X_{NN}) \in \{(0,0,1),(1,1,0)\} = A$$

in which case

$$P\{(\tilde{X}_{1N},\tilde{X}_{N1},\tilde{X}_{NN}) = (0,0,1) \mid (X_{1N},X_{N1},X_{NN}) \in A\}$$

$$= 1 - P\{(\tilde{X}_{1N},\tilde{X}_{N1},\tilde{X}_{NN}) = (1,1,0) \mid (X_{1N},X_{N1},X_{NN}) \in A\}$$

$$= \frac{1}{1+\alpha_1\beta_1} \,.$$

Then (X_{1N},X_{N1},X_{NN}) and $(\tilde{X}_{1N},\tilde{X}_{N1},\tilde{X}_{NN})$ are identically distributed since

$$P\{(\tilde{X}_{1N},\tilde{X}_{N1},\tilde{X}_{NN}) = (0,0,1)\} =$$

$$\frac{1}{1+\alpha_1\beta_1}\cdot\left(\frac{\alpha_N\beta_N+\alpha_1\beta_N\alpha_N\beta_1}{(1+\alpha_1\beta_N)(1+\alpha_N\beta_1)(1+\alpha_N\beta_N)}\right) = \frac{\alpha_N\beta_N}{(1+\alpha_1\beta_N)(1+\alpha_N\beta_1)(1+\alpha_N\beta_N)}$$

and similarly for (1,1,0).

Define everywhere else $\tilde{X}_{ij} = X_{ij}$ with all these independent and independent of the foregoing with

$$P\{X_{ij}=1\} = P\{X_{ij}=0\} = \frac{\alpha_i\beta_j}{1+\alpha_i\beta_j}$$

Let now as in case i) T_N (resp. $\tilde{T}_N$) be the row-and column sums obtained from $(s_1,\ldots,s_n;r_1,\ldots,r_n)$ (resp. $(s_1',\ldots,s_n';r_1',\ldots,r_n')$ and the X_{ij} (resp. $\tilde{X}_{ij}$) matrices.

Then

$$S_{2N} = \tilde{S}_{2N},\ldots,S_{NN} = \tilde{S}_{NN}$$

and

$$R_{2N} = \widetilde{R}_{2N}, \ldots, R_{NN} = \widetilde{R}_{NN}$$

whereas

$$S_{1N} - \widetilde{S}_{1N} = R_{1N} - \widetilde{R}_{1N} = D_N$$

again is, watched at jump times, a symmetric, recurrent random walk. Again, let us calculate the probability of a jump at stage N:

$$P\{D_N \neq D_{N-1}\} = P\{X_{1N} \neq \widetilde{X}_{1N}\}$$

$$= 2 \cdot \frac{\alpha_1 \beta_1 \alpha_N \beta_N}{(1 + \alpha_1 \beta_N)(1 + \alpha_N \beta_1)(1 + \alpha_N \beta_N)} .$$

Using again the inequality (7.1), we see that condition C_2 ensures the series to be divergent, implying that we can let

$$\tau = \inf \{N : D_N = 0\}$$

and

$$T_N' = \begin{cases} \widetilde{T}_N & \text{if } \tau \geqq N \\ T_N & \text{if } \tau < N \end{cases}$$

whereby (T_N, T_N') is a successful coupling. This completes the proof. □

We strongly suspect condition C_2 to be necessary as well but do not know how to prove this.

Asymptotic theory for the Rasch model is usually based on the assumption that $\alpha_i \in I$, $\beta_j \in J$, where I and J are compact intervals. This assumption trivially implies C_2. It would be interesting to see asymptotic theory under the assumption C_2 which, since it ensures extremality, at least implies that all the ratios $\frac{\alpha_i}{\alpha_k}$ and $\frac{\beta_j}{\beta_r}$ are consistently estimable.

10. Bibliographical Notes

The notion of a repetitive structure was introduced and investigated by Martin-Löf (1974) and it is not particularly surprising that this theory can be generalised to Polish spaces, since it relies heavily on the application of Choquet's theorem, and the Polish spaces exactly are those that can occur as sets of all extreme points of compact, convex metrisable sets.

The notion of a projective statistical field was introduced by the author (1976) where the generalisation of sufficiency etc. also was described although no results concerning the existence of projective limits etc. were discussed.

Scheffer (1971) investigated the existence of the projective limit of compact spaces and Markov kernels. Johan Vestergaard (personal communication) has pointed out a fatal error in the proof of this existence, and the paper by Lindenstrauss, Olsen and Sternfeld (1978) seems to contain a counterexample to Scheffer's result.

Scheffer's result implies namely that all projective limits of Bauer simplices (simplices with the set of extreme points being compact) are Bauer simplices, whereas Lindenstrauss et.al. show how to construct the Poulsen simplex, where the extreme points are dense, as a limit of finite-dimensional simplices. See also Poulsen (1961).

Several versions of Proposition 1.1, identifying the maximal family as a non-compact simplex, have appeared. Similarly the main theorem 4.1 has appeared in different disguises. See Accardi and Pistone

(1982), Aldous (1985), Diaconis and Freedman (1984), Dynkin (1978), Martin-Löf (1974) and Winkler (1985) for a wide range of examples, each with a slightly different emphasis and flavour.

In the setting given here, many concrete problems have been solved. The reader is referred to Dawid (1978) and Aldous (1979) for details. Salminen (1981) found the space-time harmonic functions for the Wiener-,Poisson-,Gamma-, Cauchy- and d-dimensional Bessel processes, results that all could be 'translated' to results about extremal families and canonical projective statistical fields.

Küchler and Lauritzen (1986) contains the first four of these examples and many others.

LITERATURE

Abrahamse, A.F. (1970). A comparison between the Martin boundary theory and the theory of likelihood ratios. Ann. Math. Statist. 41, 1064-1067.

Accardi, L. and Pistone, G. (1982). de Finetti's theorem, sufficiency and Dobrushin's theory. In Exchangeability in Probability and Statistics, G. Koch and F. Spizzichino (eds). North-Holland, Amsterdam.

Aldous, D.J. (1981). Representations for partially exchangeable arrays of random variables. J. Mult. Anal., 11, 581-598.

Aldous, D. (1985). Exchangeability and related topics. In Ecole d'Eté de Probabilités de Saint-Flour XIII-1983 (ed. P.L. Hennequin). Springer Lecture Notes in Mathematics, 1117. Springer, Berlin.

Aldous, D.J. and J.W. Pitman (1979). On the zero-one law for exchangeable events. Annals of Probability, 7, 704-723.

Andersen, E.B. (1970). Sufficiency and exponential families for discrete sample spaces. J. Amer. Statist. Assoc., 65, 1248-1255.

Andersen, E.B. (1973). Conditional Inference and Models for Measuring. Sc. D. Thesis, Mentalhygieinisk Forskningsinstitut, Copenhagen.

Asmussen, S. (1978). Measure-Theoretic Foundations of Probability Theory in Polish Spaces. Institute of Mathematical Statistics, Univ. of Copenhagen.

Bahadur, R.R. (1954). Sufficiency and statistical decision functions. Ann. Math. Statist., 25, 423-462.

Barndorff-Nielsen, O. (1970). Exponential Families. Exact Theory. Var. Publ. Series No. 19, Aarhus University.

Barndorff-Nielsen, O. (1978). Information and Exponential Families and conditioning. Wiley, New York.

Basawa, I.V. and Scott, D.J. (1983). Asymptotic Optimal Inference for Non-ergodic Models, Springer Lecture Notes in Statistics, **17**. Springer-Verlag.

Berg C., Christensen, J.P.R. and Ressel, P. (1984). Harmonic Analysis on Semigroups. Springer, Berlin.

Billingsley, P. (1968). Convergence of Probability Measures. Wiley, New York.

Bühlmann, H. (1960). Austauschbare stochastische Variabeln und ihre Grenzwertsätze. Univ. California Publ. Statist., **3**, 1-36.

Chentsov, N.N. (1972). Statistical Decision Rules and Optimal Conclusions. (in Russian) Nauka. Moscow.

Clifford, A.H. and G.B. Preston (1961). The Algebraic Theory of Semigroups. Vol. I., Amer. Math. Soc.

Cohn, H. (1974). On the tail events of a Markov Chain Z.f. Wahrscheinlichkeitstheorie und verw. Gebiete, **29**, 65-72.

Cohn, H. (1976). On the tail σ-field of the countable Markov chains Rev. Roum. Math. Pures et Appl. **21**, 667-675.

Cramér, H. (1946). Mathematical Methods of Statistics. Princeton Univ. Press.

Dawid, A.P. (1977). Spherical matrix distributions and a multivariate model J. Roy. Statist. Soc., B, **39**, 254-261.

Dawid, A.P. (1978). Extendibility of spherical matrix distributions. J. Multivariate Anal., 8, 567-572.

Dawid, A.P. (1982). Intersubjective statistical models. In Exchangeability in Probability and Statistics, G. Koch and F. Spizzichino (eds). North-Holland, Amsterdam.

Diaconis, P. (1977). Finite forms of de Finetti's theorem of exchangeability. Synthese, 36, 271-281.

Diaconis, P. and Freedman, D. (1980). de Finetti's theorem for Markov Chains. Annals of Probability, 8, 115-130.

Diaconis, P. and Freedman, D. (1980a). Finite exchangeable sequences. Ann. Probability, 8, 745-764.

Diaconis, P. and Freedman, D.A. (1980b). de Finetti's generalisations of exchangeability. Studies in inductive logic and probability. R. Jeffreys ed. Univ. of California Press, Berkeley.

Diaconis, P. and Freedman, D. (1981). On the statistics of vision: the Julesz conjecture. J. Math. Psychology, 24, 112-138.

Diaconis, P. and Freedman, D. (1984). Partial exchangeability and sufficiency. Proceedings of the Indian Statistical Institute Golden Jubilee International Conference on "Statistics: Applications and New Directions". Calcutta 1981. Sankhya A.

Diaconis, P. and Freedman, D. (1987a). A dozen de Finetti-style results in search of a theory. Ann. Inst. Henri Poincaré, 23, 397-423.

Diaconis, P. and Freedman, D. (1987b). A finite version of de Finetti's theorem for exponential families, with uniform asymptotic estimates. Tech. Rep. No. 91. Dept. of Statistics, Berkeley.

Diaconis, P., M.L. Eaton and S.L. Lauritzen. (1987). Applications of finite de Finetti style theorems to linear models. Tech. Rep. No. 502, School of Statistics, Univ. Minnesota.

Dynkin, E.B. (1978). Sufficient statistics and extreme points. Ann. Probability, 6, 705-730.

Feller, W. (1971). An Introduction to Probability and Its Applications. Vol. 2., Wiley, New York.

Fenstad, J.E. (1967). Representations of probabilities defined on first order languages. In Sets, Models and Recursion Theory, Ed. Crossley. North-Holland, Amsterdam.

Fenstad, J.E. (1968). The structure of logical probabilities. Synthese, 18, 1-23.

de Finetti, B. (1931). Funzione caratteristica di un fenomene aleatorio. Atti della R. Academia Nazionale del Lincei, Ser. 6. Memorie, Classe di Scienze Fisiche, Mathematiche e Naturale, 4, 251-299.

de Finetti, B. (1937). La prévision: ses lois logiques, ses sources subjectives. Ann. Inst. H. Poincaré, 7, 1-68.

Freedman, D.A. (1962). Invariants under mixing which generalize de Finetti's theorem. Ann. Math. Statist., 33, 916-923.

Freedman, D.A. (1963). Invariants under mixing which generalize de Finetti's theorem. Continuous time paramter. Ann. Math. Statist., 34, 1194-1216.

Freedman, D. and Diaconis, P. (1982). de Finetti's theorem for symmetric location families. Ann. Statist., 10, 184-189.

Föllmer, H. (1975a). On the potential theorey of random fields. Proceedings of the 40th session of ISI, Warsaw. ISI Bulletin, 46, Book 2, 362-370.

Föllmer, H. (1975b). Phase transition and Martin Boundary. Sem. Probabilités de Strasbourg IX. Springer.

Georgii, H.O. (1979). Canonical Gibbs Measures. Lecture notes in Mathematics, 760. Springer. Berlin, Heidelberg, New York.

Griffeath, D. (1975). A maximal coupling of Markov chains. Z. Wahrscheinlichkeitstheorie u. verw. Gebiete, **31**, 95-106.

Hald, A. (1960). The compound hypergeometric distribution and a system of single sampling inspection plans bases on prior distributions and costs. Technometrics, **2**, 275-340.

Hewitt, E. and Savage, L.J. (1955). Symmetric measures on Cartesian products, Trans. Amer. Math. Soc., **80**, 470-501.

Hewitt, E. and Zuckerman, H.S. (1956). The ℓ_1-algebra of a commutive semigroup, Trans. Amer. Math. Soc., **83**, 70-97.

Höglund, T.H. (1974). Central limit theorems and statistical inference for Markov chains, Z. f. Wahrscheinlichkeitstheorie u. verw. Gebiete, **29**, 123-151.

Jensen, J.L. and Johannson, B. (1988). The extremal family generated by the Yule process. TRITA-MAT-1988-1. Roy. Inst. Techn. Stockholm.

Johansen, S. (1967). Extremalpunktsmetoder i sandsynlighedsregningen. Guldmedaljeafhandling. Københavns Universitet.

Johansen, S. (1977). Homomorphisms and general exponential families, In Barra, Brodean, Romier and Van Cutsem: Recent Developments in Statistics, North-Holland. Amsterdam.

Johansen, S. (1979). Introduction to the Theory of Regular exponential families. Lect. notes 3. Univ. Copenhagen, Inst. Math. Stat.

Kallenberg, O. (1973). Canonical representations and convergence criteria for processes with interchangeable increments. Z. Wahrscheinlichkeitstheorie u. verw. Gebiete, **27**, 23-36.

Keiding, N. (1974). Estimation in the birth process, Biometrika, **61**, 71-80.

Kingman, J.F.C. (1978). Random partitions in population genetics. Proc. R. Soc. Lond. A, **361**, 1-20.

Koch, G. and Spizzichino, F. (eds). (1982). Exchangeability in Probability and Statistics. North-Holland, Amsterdam.

Küchler, U. and Lauritzen, S.L. (1986). Exponential families, extreme point models and minimal space-time invariant functions for stochastic processes with stationary and independent increments. Res. Rep. R-86-18, Inst. Electr. Syst., Aalborg Univ.

Lauritzen, S.L. (1972). Sufficiency and time series analysis, Univ. Copenhagen. Inst. Math. Stat., Preprint No. 11.

Lauritzen, S.L. (1973). Maximum likelihood prediction, Mimeographed. Seminar Notes. Univ. Copenhagen, Inst. Math. Stat.

Lauritzen, S.L. (1974). On the interrelationships among sufficiency and some related concepts, Univ. Copenhagen, Inst. Math. Stat., Preprint No. 8.

Lauritzen, S.L. (1974). Sufficiency, Prediction and extreme models, Scand. J. Statist., 1, 128-134.

Lauritzen, S.L. (1975). General exponential models for discrete observations, Scand. J. Statist., 2, 23-33.

Lauritzen, S.L. (1976). Projective Statistical Fields, Univ. Copenhagen, Inst. Math. Stat., Preprint No. 6.

Lauritzen, S.L. (1984). Extreme point models in statistics. Scand. J. Statist., 11, 65-91.

Lindenstrauss, J., Olsen, G. and Sternfeld., Y. (1978). The Poulsen simplex. Ann. Inst. Fourier Grenoble, 28, 91-114.

Martin-Löf, P. (1970). Statistiska Modeller. Notes by Rolf Sundberg Mimeographed.

Martin-Löf, P. (1974). Repetitive structures. In Proceedings of Conference of Foundational Questions in Statistical Inference, Aarhus 1974. Memoirs 1. Ed. O. Barndorff-Nielsen, P. Blæsild and G. Schou.

Martin-Löf, P. (1975). Reply to Sverdrup's polemical article: tests without power. Scand. J. Statist., 2, 161-165.

Neveu, J. (1964). Chaines de Markov et théorie du potentiel. Ann. Faculté des Sciences Clermont, 3, 37-65.

Orey, S. (1971). Limit Theorems for Markov Chain Probabilities. Van Nostrand. London.

Petrich, M. (1977). Lectures in Semigroups, Wiley, Akademie-Verlag.

Phelps, R.R. (1966). Lectures on Choquet's Theorem. Van Nostrand, Princeton, N.J.

Pitman, J.W. (1978). An extension of de Finetti's theorem. Adv. Appl. Prob., **10**, 268-269.

Poulsen, E.T. (1961). A simplex with dense extreme points. Ann. Inst. Fourier Grenoble, 11, 83-87.

Rasch, G. (1960). Probabilistic Models for Some Intelligence and Attainment Tests. Danmarks Pædagogiske Institut. Studies in Mathematical Psychology I.

Ressel, P. (1985). de Finetti type theorems: An analytical approach. Ann. Prob., **13**, 898-922.

Ressel, P. (1986). Integral representations of symmetric stochastic processes. Manuscript.

Salminen, P. (1981). Martin boundaries for some space-time Markov processes. Z. Wahrscheinlichkeitstheorie u. verw. Gebiete, **55**, 41-53.

Scheffer, C.L. (1971). La limite projective des espaces topologiques de Hausdorff compacts dans la catégorie des diffusions continues sousmarkoviennes. Co. R. Acad. Sc. Paris, **272**, 1198-1201.

Schoenberg, I.J. (1938). Metric spaces and positive definite functions. Trans. Amer. Math. Soc., **44**, 522-536.

Skibinsky, M. (1967). Adequate subfields and sufficiency. Ann. Math. Statist., **38**, 155-161.

Smith, A.F.M. (1981). On random sequences with centered spherical symmetry. J. Roy. Statist. Soc. Ser. B., **43**, 208-209.

Spitzer, F. (1976). Principles of Random Walk, 2nd ed. Springer. New York, Heidelberg. Berlin.

Sweeting, T. (1986). Asymptotic conditional inference for the offspring mean of a supercritical Galton-Watson process. Ann. Statist., 14, 925-933.

Wilson, A.G. (1967). A statistical theory of spatial distribution models. Transportation Research, 1, 253-269.

Winkler, G. (1985). Choquet Order and Simplices with Applications in Probabilistic Models. Lecture Notes in Mathematics, No. 1145, Springer, Berlin.

Zaman, A. (1984). Urn Models for Markov exchangeability. Ann. Probability, 12, 223-229.